Analog Switches
Applications & Projects

Analog Switches
Applications & Projects
Delton T. Horn

TAB BOOKS
Blue Ridge Summit, PA

FIRST EDITION
FIRST PRINTING

Copyright © 1990 by TAB BOOKS
Printed in the United States of America

Library of Congress Cataloging-in-Publication Data

Horn, Delton T.
    Analog switches : applications & projects / by Delton T. Horn.
       p.   cm.
    ISBN 0-8306-8445-X   ISBN 0-8306-3445-2 (pbk.)
    1. Switching circuits.   2. Analog electronic systems.   I. Title.
TK7868.S9H67   1990
621.381'5372—dc20   89-48456
                                                         CIP

TAB BOOKS offers software for sale. For information and a catalog, please contact
TAB Software Department, Blue Ridge Summit, PA 17294-0850.

Questions regarding the content of this book should be addressed to:

Reader Inquiry Branch
TAB BOOKS
Blue Ridge Summit, PA 17294-0214

Acquisitions Editor: Roland S. Phelps
Book Editor: Christopher M. Cooke
Production: Katherine Brown

Contents

Introduction

The most complex electronic systems are made up of a number of simpler sub-circuits. A handful of basic circuit types, such as oscillators and amplifiers, are used time after time in countless electronic systems.

One of the most important, but frequently neglected, types of basic circuits is the *switching circuit*. This circuit is used to route electrical signals from one point in an electronic system to another. Almost every advanced electronics circuit incorporates some kind of switching.

Switching devices range from simple mechanical devices through advanced digitally controlled switching networks. This book covers the entire range of switching circuitry, from the simple to the sophisticated.

To get the most out of this book, you should have some prior experience with electronics, at least at the hobbyist level. The material presented here is not advanced, and the math is kept to a minimum. You do not have to be an advanced electronics professional, but you should have a firm grasp on basic electronics principles.

This book is aimed more at the hobbyist than the professional technician, although even professionals will find some useful information here.

The emphasis in this book is on theory and how the various switching circuits work. There are also a few practical circuits included for you to breadboard and experiment with.

Although short, this book covers quite a bit of ground. There are a great many different approaches to electronics switching circuits. Switching is not as simple a topic as it might seem.

1
Basics of Switching

SWITCHING IS PROBABLY THE MOST BASIC ELECTRONIC FUNCTION THERE IS. The simplest kinds of switches are mechanical devices with connectors on *sliders* that physically move between one or more sets of contacts.

In its simplest form, a switch has two contacts, or leads, which are connected to separate points in a circuit. The slider in such a simple switch has two possible positions. Either the slider is positioned so that there is an electrical connection between two contacts, or it is positioned so that the electrical path between the two contacts is broken.

If the contacts are electrically connected, the switch is said to be *closed*. When the slider is moved to break the electrical connection between the contacts, the switch is said to be *open*. In many applications, a closed switch is "on," and an open switch is "off."

As you can see, the basic switch is a very simple device. It might seem that I already have covered everything there is to say about the simple switch. In a sense, I have, but there are countless variations on this simple electronic device. They will be discussed and examined throughout this book. The emphasis will be on switching in analog circuits. Digital switching circuits will not be covered here.

The most familiar kinds of switches are mechanical devices, like the basic switch just discussed. Mechanical switches will be discussed in detail in this chapter. Later chapters will explore electronic switching circuits.

SWITCH CONFIGURATIONS

One important way switches differ is in the number and arrangement of their contacts. A switch must have at least two contacts, and can have more than two contacts for controlling more than a single circuit or subcircuit.

Four common switch patterns are used widely in mechanical switches. (These switching patterns also are simulated frequently in electronic switching circuits.) The four common switching patterns are known by the following names and acronyms:

SPST	Single Pole Single Throw
SPDT	Single Pole Double Throw
DPST	Double Pole Single Throw
DPDT	Double Pole Double Throw

Pole is essentially another term for slider. A *single pole* switch has just one slider, as in the basic switch described earlier in this chapter. A *double pole* switch has two *ganged* (or connected) sliders, which move together, but control different sets of contacts.

The term *throw* indicates how many closed positions the switch has. A *single throw* switch has one closed position, and one open position. A *double throw* switch, on the other hand, has two closed positions and no open positions. One contact, called the *common* contact, is always electrically connected to the slider. When the slider is in one position, the common contact is electrically shorted to a second contact. Moving the slider to its alternate position isolates the common contact from the second contact, but connects it electrically to a third contact. One set of contacts or the other is activated (closed) at any given time, depending on the position of the switch's slider.

It is possible for a switch to have more than two sliders and/or more than two closed positions. Such complex mechanical switches will be discussed later in this chapter, in the section "Rotary Switches."

Each of the basic switch types has a different number of contacts, as follows:

Type	Contacts
SPST	2
SPDT	3
DPST	4
DPDT	6

A DPDT switch can be used in place of any of the other three basic switch types. A SPDT or a DPST switch can be substituted for a SPST switch. In each of the substitutions, the extra contacts on the replacement switch are simply ignored and left disconnected from the circuit.

The SPST Switch

The operation of each of the basic switch types is made clearer by their schematic symbols. The schematic symbol for a single pole single throw switch is shown in FIG. 1-1. Notice that the two contacts, or circuit connection points, are clearly indicated.

Fig. 1-1. A SPST switch controls a single circuit.

If the slider (the small arrow in the schematic symbol) is in the open position, there is an open circuit between the contacts. Current cannot flow between the contacts in this situation. The effect will be the same as if the switch was removed from the circuit.

If the slider is in it closed positions, there will be a short circuit between the switch's contacts, and current will flow between them, just as if the switch was an ordinary length of wire.

Notice that it is customary to show a SPST switch with the slider in the open position in most schematic diagrams. This makes the symbol easier to see at a glance. FIGURE 1-2 shows the symbol for a closed SPST switch.

The direction of the slider arrow for the symbol for a SPST switch is usually irrelevant. Draw the slider in whichever direction looks better.

Fig. 1-2. A closed SPST would be schematically indicated with this symbol.

The SPDT Switch

The next basic switch type is the single pole double throw switch. The schematic symbol for this device is illustrated in FIG. 1-3. Notice that there are three contacts. The common contact is at the base of the slider arrow. Because there is no fully open position, the slider of a SPDT switch is shown in one of its two closed positions.

The choice of the indicated slider position in the schematic diagram will depend on the specific circuit being drawn. If there is no clear cut "normal" position, the slider may be shown in either position.

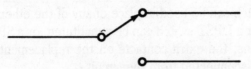

Fig. 1-3. A SPDT switch connects a common circuit with either of two other circuits.

In effect, the SPDT switch is like two SPST switches, connected as shown in the circuit of FIG. 1-4. The sliders of these two switches are ganged together. (This is indicated by the dotted line between the sliders.) When one switch is open, the other one is closed, and vice versa.

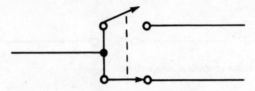

Fig. 1-4. A SPDT is equivalent to a pair of ganged SPST switches.

The DPST Switch

The double pole single throw switch is really two SPST switches with a common slider. This is clearly indicated in the schematic symbol for this type of switch, which appears in FIG. 1-5. The dotted line between the two sliders indicates that they are ganged, or tied together. When one is moved, the other must move in the same direction. Either both switch sections are open, or both are closed. There are no other possible combinations with the DPST switch.

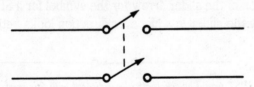

Fig. 1-5. A DPST switch is like two ganged SPST switches in parallel.

In some schematic diagrams, the two switch sections may be drawn widely separated. If there is a dotted line connecting the sliders of two (or more) switch units in a schematic diagram, a single switching device is indicated, and the sliders must work in unison at all times.

Of the four basic switch configurations (SPST, SPDT, DPST, and DPDT), the DPST switch is by far, the least commonly used. DPST switches can be difficult to find on the hobbyist market. When a DPST function is needed in a circuit, a DPDT switch is normally used, with two of the switch contacts left disconnected.

The DPDT Switch

The double pole double throw switch is the most sophisticated and versatile of the four basic switch configurations (SPST, SPDT, DPST, and DPDT). As implied by the schematic symbol, shown in FIG. 1-6, the DPDT is the equivalent of two SPDT switches with ganged sliders. When one slider is moved, the other will automatically be moved too, and in the same direction.

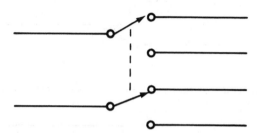

Fig. 1-6. The most sophisticated and versatile of the four basic switch configurations is the DPDT switch.

In some schematic diagrams, the two switch sections may be drawn widely separated. If there is a dotted line connecting the sliders of two (or more) switch units in a schematic diagram, a single switching device is indicated, and the sliders must work in unison.

A DPDT switch can be used to duplicate the function of any of the other basic switch configurations (SPST, SPDT, or DPST). It is just a matter of determining which of the six available contacts to use. The unneeded contacts are left disconnected from the circuit.

Other Switch Configurations

Other switch configurations with more than two sliders (poles) and/or more than two positions (throws) are also possible. Usually, a number greater than two is indicated by a numeral rather than a letter in the switch identification. For example, FIG. 1-7 shows a 3PDT (three pole double throw) switch. There are three ganged sliders (or poles), each with two possible positions—a common contact and one or the other of two end contacts. A 3PDT switch has a total of nine contacts.

FIGURE 1-8 illustrates a SP4T (single pole four throw) switch. This type of switch has just a single slider, but it can be set for any of four possible positions. With the switch shown here, there are five contacts.

Many other switch configurations are also used. More information on multiple position switches will be given in the section on rotary switches, later in this chapter.

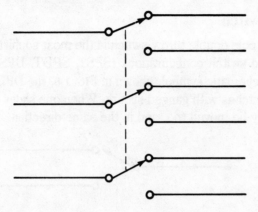

Fig. 1-7. Nonstandard switch configurations, such as a 3PDT switch, are also possible.

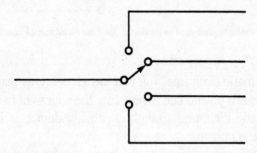

Fig. 1-8. A SP4T switch is another example of an uncommon switch configuration.

MANUAL SWITCH TYPES

A number of different mechanisms can be used to construct a manual switch. In this section we will look at some of the most common types of manual switches. Each switch type has its own advantages and disadvantages, and is suitable for a different set of applications.

Knife Switches

One of the simplest types of manual switches is the *knife switch*, which is illustrated in FIG. 1-9. The handle of the knife switch is metal and can conduct current. When the switch handle is in the position shown in FIG. 1-9A, no current can flow through the attached circuit because there is not a complete path for the current to flow between the contacts.

When the switch handle is moved to the position illustrated in FIG. 1-9B, the metal handle creates a short circuit between the two contacts. The circuit path is completed (closed), and current can flow through the circuit.

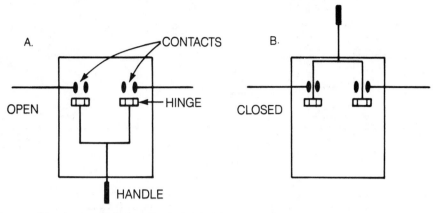

Fig. 1-9. The simplest type of mechanical switch is the knife switch.

A knife switch is very simple and inexpensive. Home made knife switches can be easily constructed. Another advantage of this type of switch is that it is very easy to tell at a glance if the switch is open or closed.

The knife switch also has many important disadvantages. It is bulky, the conducting portion of the switch is fully exposed, (creating a potential shock hazard) and contaminents can easily corrode the metallic handle and the switch contacts. Knife switches are used rarely in modern electronic circuits.

Slide Switches

Probably the most commonly available and widely used type of manual switch is the *slide switch*. Slide switches are relatively inexpensive and safe (all electrical conductors are enclosed within the body of the switch). They are also reasonably reliable, due to their simple mechanical construction.

A knob is used to move the slider (movable portion) from position to position. A metallic strip is mounted on the base of the slider to make an electrical connection between the contacts when in the correct position. The slide switch is so popular, the term "slider" is used to describe the moving portion of most mechanical switches, even though the term really doesn't make much sense in some cases.

FIGURE 1-10 shows the basic construction of a SPST slide switch. When the slider is in the position shown in FIG. 1-10A, the circuit is open and no current can flow between the switch's contacts. But, when the slider is moved into the position illustrated in FIG. 1-10B, the metal strip along the bottom of the slider touches the two contacts, making an electrical connection that completes the current path in the circuit.

A SPDT slide switch is quite similar to the SPST switch of FIG. 1-10. This type of switch is illustrated in FIG. 1-11. The only difference here is the addition

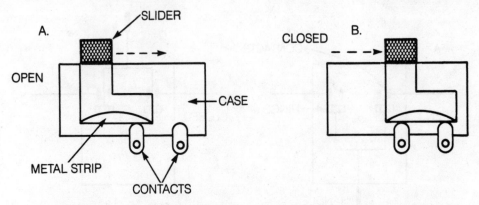

Fig. 1-10. The internal construction of a SPST slide switch is quite simple and straightforward.

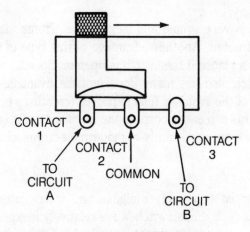

Fig. 1-11. A SPDT slide switch is very similar to a SPST slide switch.

of the third contact. The slider's metal strip always creates a short circuit between two (and only two) of the switch's contacts. Either the metal strip connects contacts 1 and 2 (common), or contacts 2 (common) and 3.

Some SPDT switches are designed to have a center-off position, as illustrated in FIG. 1-12. In the center position, the slider is touching the common (#2) contact, but neither of the end contacts (#1 or #3), so there is no complete current path.

A DPST switch is like two side-by-side SPST slide switches. A bottom view of a DPST slide switch appears in FIG. 1-13. Notice that the slider has two separate metal strips along its bottom side. Each strip controls a different set of contacts on the switch. A DPDT slide switch is like a pair of side-by-side SPDT slide switches. This type of switch is shown in FIG. 1-14. DPDT slide switches with a center-off position are available also.

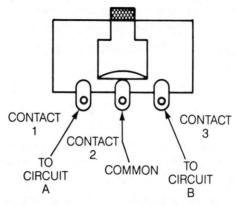

Fig. 1-12. Some SPDT slide switches have a center off position.

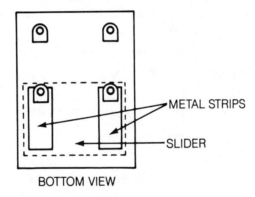

BOTTOM VIEW

Fig. 1-13. A DPST slide switch is like two side-by-side SPST slide switches in a single housing.

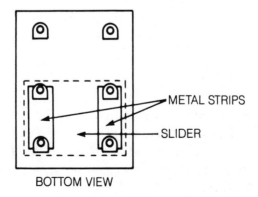

BOTTOM VIEW

Fig. 1-14. A DPDT slide switch is like two side-by-side SPDT slide switches in a single housing.

The chief disadvantage of slide switches, especially for electronics hobbyists, is physical shape. Slide switches are almost always square, or rectangular. It is difficult for many hobbyists to drill or punch out a neat rectangular hole in a project case.

Toggle Switches

Probably the second most popular type of switch is the toggle switch. The toggle switch is quite similar in concept to the slide switch. This is indicated clearly in FIG. 1-15, where the internal construction of a typical toggle switch is shown. The primary difference here is that the movable slider is in the shape of a ball that rolls in and out of position as the handle is moved. Toggle switches are also available in SPDT, DPST, and DPDT configurations.

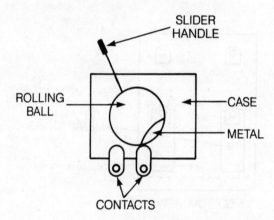

Fig. 1-15. The toggle switch is similar in concept to the slide switch.

Toggle switches usually are manufactured to fit in round holes for mounting. This makes things very easy for the electronics hobbyist drilling a customized case or control panel for a project. The main disadvantage of the toggle switch is that it is relatively expensive, compared to a similar slide switch.

Pushbutton Switches

Another popular type of switch is the pushbutton switch. A pad, or button, is depressed to move the slider into position towards, or away from the switch's electrical contacts. Most pushbutton switches are of the momentary action type (discussed further in the next section of this chapter.) Pushbutton switches are usually of the SPST type, although they occasionally are available in other configurations (SPDT, DPST, and DPDT). Like the toggle switch, this device fits neatly into an easily drilled hole.

Some pushbutton switches automatically latch the slider in one position or the other. This type of switch often is called a "push-on/push-off" switch. Assuming SPST contacts, the first time the button is pushed, the switch contacts will be closed (the circuit will be turned on). Pushing the button a second time will open the switch contacts (the circuit will be turned off). Either position can be held indefinitely, until the button is pushed again.

There is a great deal of variation in the quality of pushbutton switches. Some switches of this type are fairly inexpensive, while others may cost several dollars apiece. The low-cost pushbutton switches tend to have a rather limited lifespan, and are prone to getting stuck. Still, in many low-demand applications, an inexpensive pushbutton switch may do the job just fine, and there would be no point in investing in a more costly unit.

Momentary Action Switches

Most switches latch into one position or the other. When the operator moves the slider, it stays put, until the operator moves it again. A circuit can be turned on or off.

In some applications, you only want the slider to hold one position very briefly. You could quickly flick the switch back and forth, but that would be a highly awkward solution at best. For such applications, use a momentary action switch. A momentary action switch has a spring loaded slider, which automatically returns to a specific rest position whenever it is released.

The vast majority of momentary action switches are designed for SPST operation and there are two varieties of momentary action SPST switches. Their difference lies in which slider position is the normal rest position. One type is *normally open*, the other is *normally closed*.

A normally open (or N.O.) switch usually has the slider positioned away from the switch's electrical contacts. When the switch is left alone, the contacts are open, and the controlled circuit is off. When you activate the switch, the contacts are closed (and the circuit is turned on) for as long as you hold the switch in place. As soon as you release the switch it immediately reverts to its normal rest state, open in this case.

The normally closed (or N.C.) switch works in just the opposite manner as a normally open switch. This type of switch has its slider positioned to short out the switch's electrical contacts. When the switch is left alone the contacts are closed and the controlled circuit is on. When you activate the switch, the contacts are opened (and the circuit is turned off) for as long as you hold the switch in place. As soon as you release the switch it immediately reverts to its normal rest state, and the switch contacts are closed.

Momentary action switches are usually of the pushbutton type. Spring-loaded momentary action slide and toggle switches are available also, but they are relatively rare.

For convenience, momentary action switches normally are shown as push-button switches in schematic diagrams. FIGURE 1-16 shows the schematic symbol for a normally open SPST switch. The schematic symbol for a normally closed SPST switch appears in FIG. 1-17. Notice in each case that the switch is shown in its rest position.

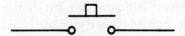

Fig. 1-16. The schematic symbol for a Normally Open SPST push-button switch.

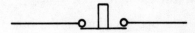

Fig. 1-17. The schematic symbol for a Normally Closed SPST push-button switch.

Inexpensive momentary action switches tend to wear out quickly and are often prone to jamming. If a switch failure would be a problem in your specific application, it is well worthwhile to invest the extra expense of high-quality switches.

Mercury Switches   An unusual type of momentary contact switch is the *mercury switch*. This device is used generally in motion detecting applications, such as the tilt switch in pinball machines.

The basic construction of a mercury switch is illustrated in FIG. 1-18. Two separated electrodes are enclosed in a glass tube. The tube also contains a glob-ule of mercury. Mercury is a metal which is in liquid form at room temperature.

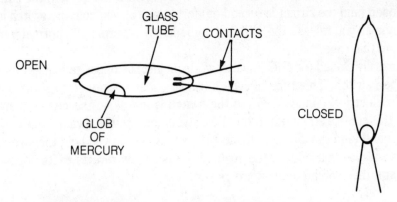

Fig. 1-18. A mercury switch is often used for motion detection.

As the glass tube is moved, the globule of mercury moves about within it. If the tube is held in the proper position, the mercury will flow over both of the internal electrodes, causing a short circuit between them.

When the mercury is in position to short out the two electrodes, the switch is closed. Otherwise, the switch is open, and current cannot flow between the two contacts. Mercury switches are usually of the SPST configuration.

Magnetic Reed Switches Another specialized type of momentary action switch is the *magnetic reed switch*. This type of switch finds its most frequent application in intrusion alarm systems. Magnetic reed switches are almost always of the SPST configuration.

A reed switch consists of two sections, as shown in FIG. 1-19. One section (A) contains the actual switching mechanism, while the other (B) contains a small permanent magnet. The switch section is made up of tiny, magnetism sensitive reeds. The presence of a sufficiently strong magnet field causes the switch contacts to leave their normal, rest position. Magnetic reed switches are available in both normally open and normally closed versions. (The choice of which to use will depend on the specific application.)

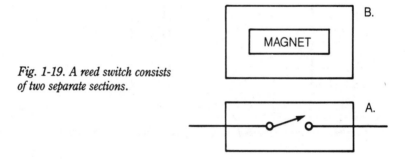

Fig. 1-19. A reed switch consists of two separate sections.

In a typical application, a reed switch is mounted on a door, as illustrated in FIG. 1-20. The magnet section is mounted on the movable portion of the door itself. The switch section is mounted to the stationary door frame, with wires leading off to the controlled circuit. The two switch sections are positioned so that when the door is closed, the magnet is as close as physically possible to the switch section. Opening the door moves the magnet away from the switch section.

If a normally open reed switch is used in FIG. 1-20, the switch will be open when the door is opened. Closing the door will cause the switch contacts to close. If a normally closed reed switch is substituted in this system the opposite will result. The switch will be closed when the door is open and closing the door will cause the switch contacts to open.

A reed switch can be mounted almost anywhere, as long as the magnet section can be moved close to and away from the switch section. For example, the

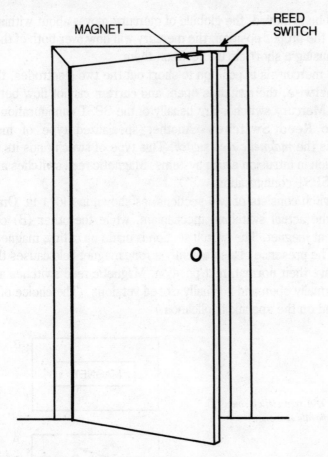

Fig. 1-20. Reed switches are often mounted on doors and windows in intrusion detector systems.

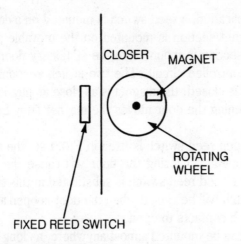

Fig. 1-21. Another application for a reed switch is as a rotation counter.

magnet section could be mounted on a rotating shaft, as shown in FIG. 1-21. The switch will be activated once per rotation, as the magnet passes by the switch section. A counting circuit and a timing circuit could be used to determine the number of rotations per minute (rpm).

Potentiometer Switches

One specialized type of switch that commonly is used is the *potentiometer switch*. As the name suggests, this is a switch which fits onto the back of a potentiometer. Potentiometer switches are usually, though not always, of the SPST configuration.

The switch is controlled by the potentiometer's shaft, or knob. When the potentiometer's shaft is at its maximum resistance position, the switch is open (off). But as soon as the control knob is advanced from this extreme position, the switch contacts click shut, turning the controlled circuit on. From then on, the potentiometer operates just like a normal unswitched potentiometer.

The potentiometer and the switch usually will be employed in the same section of the circuit, but this isn't necessarily the case. The potentiometer and the switch are tied mechanically together, but are electrically distinct. They have no internally shared electrical contacts, making them electrically independent. They could be used in two separate circuits, although the mechanical linkage probably would cause some confusion in operating the circuits.

The most common application for a potentiometer switch is to turn the main power supply of a circuit on and off. For example, an audio amplifier might have a switch connected to its master volume control potentiometer. When the volume is turned all the way down to its minimum level (maximum resistance setting), the power to the circuit is switched off.

DIP Switches

A fairly recently developed type of switch is the *DIP switch*. This type of switch commonly is used in computer equipment and other digital circuitry.

A DIP switch is actually a self-contained group of several miniaturized switches—usually SPST switches. The housing of a DIP switch is the same size as a DIP (Dual Inline Pin) IC package, and can be easily plugged into an IC socket.

A typical four SPST DIP switch unit has eight pins—two for each of the switch sections. All of the switches are independent, both electrically and physically. They do not have to be operated together. Each individual switch has its own individual slider.

Because the switches are so small, the sliders often are moved with a pencil tip or a small screwdriver, or some other similar implement. Most people's fingers are too large to operate just one of the sliders, without accidentally moving one of the other switches too.

DIP switches are almost never used for frequently operated front panel controls. They usually are employed to set customized operating parameters which are changed rarely. For example, a computer printer might use a DIP switch to set the baud rate or the print font.

Rotary Switches

Rotary switches are used for multipole and/or multiposition switching functions. Most of the switch types described so far have no more than two poles and two positions. In some circuits, more versatile switching is needed. Here is where rotary switches come in.

The term *rotary switch* comes from the fact that a rotating slider is used. It is controlled by a shaft and knob, similar to that of a potentiometer.

The schematic symbol for a rotary switch varies somewhat, depending on the number of poles and positions for the specific switch, but certain features are common. The rotating slider is indicated with the other switch contacts arranged in a semicircular pattern. As an example, FIG. 1-22 shows the schematic symbol for a SP12T (Single Pole Twelve Throw) rotary switch. FIGURE 1-23 illustrates the schematic symbol for a 3P6T (Three Pole Six Throw) rotary switch. Many other combinations are possible also.

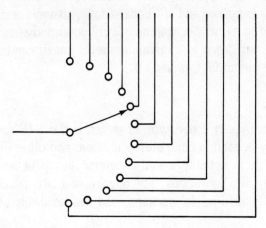

Fig. 1-22. A typical configuration for a rotary switch is a SP12T switch.

If you can't find the specific rotary switch called for in a project, you always can use a switch with too many poles or too many positions (throws). Just leave the unused contacts disconnected from the circuit. If you use a rotary switch with too many positions, your knob will have some *dead*, or unused positions.

There are two basic varieties of rotary switches: *nonshorting*, and *shorting*.

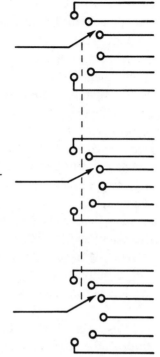

Fig. 1-23. Another typical configuration for a rotary switch is a 3P6T switch.

The nonshorting rotary switch disconnects the circuit at one position completely before the connection to the next position is made. For example, consider moving the slider from contact #1 to contact #2. At the mid-point between these two contacts, the slider will be electrically isolated from both.

The other type of rotary switch is the shorting type. Another term used for this type of switch is the *make-before-break* type rotary switch. When moving the slider from contact #1 to contact #2, the slider will make an electrical connection with both contacts for a brief instant, before the connection to contact #1 is broken.

In most circuits it probably won't matter which type of rotary switch is used, however, some specialized circuits may require one type or the other. In constructing a project from a published set of plans, pay attention to any relevant notes. If a nonshorting rotary switch is called for, do not substitute a shorting rotary switch. If the plans do not specify which type is required, you can reasonably assume it doesn't matter for that particular circuit.

RELAYS

In many practical applications, it might not be desirable to use any of the basic manual switch types discussed so far in this chapter. For instance, you might need a circuit that must be switched on when the voltage in another circuit rises above some specified level. You could sit around and watch a voltmeter connected to the second circuit and manually flick a switch in the first circuit at the appropriate moment. That would be a highly impractical, inconvenient, and inefficient solution. Some sort of automated switching would clearly be a much better choice in such an application.

An automatic switch of some sort would be necessary also in any remotely controlled circuit. Often the circuit being switched is not readily accessible. It would be very inconvenient to run a pair of wires carrying the full power supply voltage or electrical data over a long distance to a convenient control point. It would be far more efficient to just send a small control voltage over light-duty connecting wires (or possibly even radio waves) to a remote-controlled electrically automated switch.

There are many approaches to automated switching in electrical circuits. The most important of these will be covered throughout the later chapters of this book. Most of these switching circuits are purely electronic. There is one simple form of mechanical automated switching. It is a device known as a *relay*.

A relay consists of two parts—a coil and a magnetic switch. When an electrical current (dc) flows through any coil, a magnetic field will be created around it. The strength of this magnetic field is directly proportional to the amount of current flowing through the coil. In a relay, current is fed through the coil. At some specific point, the magnetic field around the coil will be strong enough to pull the switch's slider from its rest, or *de-energized*, position, to its momentary, or *energized*, position. As long as the coil current remains higher than the critical value, the relay's switch contacts will be held in their energized positions.

If the electrical power through the coil drops, so will the strength of the magnetic field surrounding it. If the current is cut off completely, the magnetic field will collapse to zero. If the magnetic field drops below the relay's critical value, the switch slider will be released, allowing it to spring back to its original, de-energized position.

The switch section of a relay may be any of the basic switching configurations discussed earlier in this chapter (SPST, SPDT, DPST, or DPDT.) For single throw units (SPST or DPST), the contacts may be either normally open or normally closed, depending on the specific requirements of the individual application. With an SPDT, or DPDT switching configuration, there is both a normally open and a normally closed set of contacts.

The schematic symbol for a SPST relay is shown in FIG. 1-24. The switch contacts are drawn in their normal (de-energized) position. The symbol used in schematic diagrams to represent a SPDT relay is illustrated in FIG. 1-25. Relays usually are identified in schematic diagrams and parts lists by the letter *K*.

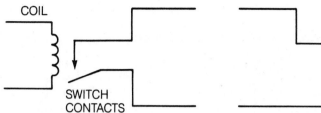

Fig. 1-24. The schematics symbol for a relay, an electrically controlled mechanical switch.

Fig. 1-25. The schematic symbol for a SPDT relay.

A relay's coil and its switching contacts almost always are used in electrically isolated circuits. That is, the current through one circuit (the coil circuit) controls the switching of another circuit.

Relays vary greatly in size, depending primarily on the amount of power they can carry safely. I've seen relays that will fit comfortably on the tip of your little finger, and relays that a grown man would have difficulty lifting. In hobbyist electronic circuits, relays are seldom more than a few inches in dimension.

Separate power ratings typically are given for a relay's coil and switch contacts, because they generally are used in separate circuits (or subcircuits). Relays range from tiny units intended for 0.5 watt (or less) transistorized equipment on up to huge megawatt (millions of watts) devices used in industrial power generating plants. At either extreme, the principle of operation is precisely the same.

The most important rating for a relay is the amount of power (usually given as a voltage) required to make the switch contacts move to their energized position. This is called *tripping* the relay. Typical trip voltages for common relays used in electronic circuits are 6, 12, 24, 48, 117, and 240 volts. Both ac and dc types are available.

The controlling voltage through a relay coil should be kept within about ± 25% of the rated value. Too large a voltage could burn out the coil windings. On the other hand, too small a control voltage could result in erratic operation of the relay.

In some applications, it may be necessary to drive a relatively high-power circuit with a fairly low-power control signal. This can be done with a circuit like the one illustrated in FIG. 1-26. High-voltage supply B is in operation *only* when the relay is energized, minimizing power consumption.

Occasionally, the available control signal will not be sufficient to drive a large enough relay for the load circuit to be controlled. In a situation like this, the solution may be to add an extra medium-power relay to act as an intermediate stage, as shown in FIG. 1-27.

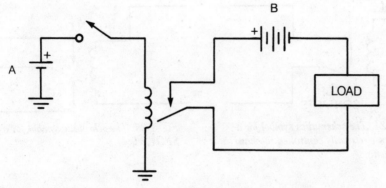

Fig. 1-26. This circuit can be used to increase the power handling capability of a relay.

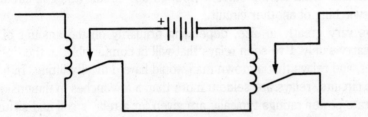

Fig. 1-27. Large loads can sometimes be controlled with a low power signal by using an intermediate relay.

The coil winding of a relay could self-destruct if the current through it changes very suddenly. One way this could happen is if a series switch, as shown in FIG. 1-26, is opened up. The voltage through the relay coil drops from V+ down to 0 in a tiny fraction of a second. This causes the magnetic field around the coil to collapse very rapidly. This abrupt change in the magnetic field will induce a brief high-voltage spike in the relay. This voltage spike could be large enough to damage or destroy the coil or the switch contacts.

In many practical circuits, a diode often is placed in parallel with the relay coil to suppress such high-voltage transients (which are known as *back EMF*). This simple, but effective precaution is illustrated in FIG. 1-28. With this arrangement, the diode limits the voltage through the relay coil to the power supply voltage (unless, of course, the back EMF spike is large enough to destroy the diode itself.)

Almost any standard diode may be used in this application. The requirements here are not at all critical. The only specification you need to be concerned with is the PIV (Peak Inverse Voltage) rating of the diode. The diode's PIV should be high enough to cover the anticipated voltage that will be applied across the device.

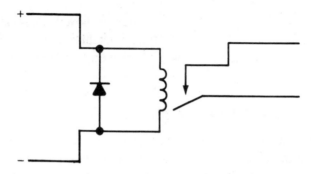

Fig. 1-28. A diode for protection from back EMF, which can damage a relay's coil during sudden current changes.

To drive high-power relays from low-power control signals, often a transistor amplifier circuit is used. A low-power battery can be used in the coil/amplifier circuit, and a higher-power supply can be available in the switched circuit. This limits undue power drain. The high-power supply is used only when it is needed. A typical amplified relay circuit of this type is illustrated in FIG. 1-29.

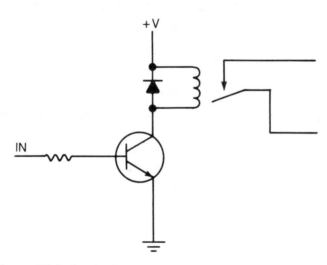

Fig. 1-29. An amplified relay circuit.

Most common relays are momentary action types. They are energized only while sufficient power is flowing through the coil. When the coil current drops, the switch contacts are released to their de-energized position.

Some relays are of a latching type. Not surprisingly, a device of this sort is called a *latching relay*. One control pulse closes the switch contacts, which will remain closed, even if the control pulse is now removed. A separate control pulse opens the switch contacts. Each time a latching relay is triggered, its switch contacts latch into the appropriate position until the next trigger signal is received.

FUSES AND CIRCUIT BREAKERS

Another kind of automatic switching is used specifically for circuit protection. The voltage through a circuit can be controlled easily by the design of the power supply, except, perhaps, for a few relatively rare (and brief) transients. The current drawn through a circuit depends on the resistance and impedance factors within the load circuit itself. If the resistance drops because of a short circuit, or some other defect, the current could rise rapidly to a level that can damage or destroy some of the components. In many cases, such excess current could present a significant shock or fire hazard.

What is needed under such circumstances is a way to disconnect the power supply from the load circuit quickly, before the excess current has a chance to do much damage. This is done most often with a specialized device called a *fuse*. A fuse is usually a thin wire that is manufactured carefully, so that it will melt and break when the current passing through it exceeds a specific value. The schematic symbol for a fuse is shown in FIG. 1-30.

Fig. 1-30. The schematic symbol for a fuse, used to protect a circuit from excessive current.

Fuses usually are enclosed in glass (or sometimes metal or ceramic) tubes for protection. The fuse wire is very thin and could be easily damaged if it was exposed. A typical fuse is shown in FIG. 1-31.

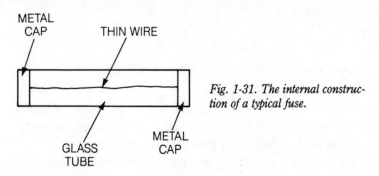

Fig. 1-31. The internal construction of a typical fuse.

FIGURE 1-32 illustrates a simplified circuit demonstrating the operation of a fuse. If the current drawn by the load circuit exceeds the current rating of the

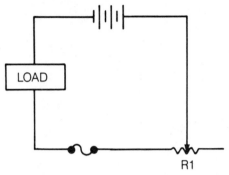

Fig. 1-32. This circuit demonstrates the functioning of a fuse.

fuse for any reason (demonstrated by changing the value of R1 in FIG. 1-32), the fuse will blow, opening the circuit. Once the fuse is blown, no further current will flow through the circuit.

Sometimes fuses are soldered directly into a circuit. But once a fuse element has been melted, it must be replaced before the circuit can be reused. For this reason, soldered fuses are rather impractical in most applications. For more convenient fuse replacement, some kind of socket generally is used for fuses. Most commonly, the fuse is held between a set of spring clips. Another frequently used method is to fit the fuse into a special receptacle with a screw cap.

Frequently replacing fuses can be a nuisance, so sometimes a component called a *circuit breaker* is used in place of a fuse. A circuit breaker is a special SPST switch that will automatically open if the current through it exceeds some specific level. To close the switch again, you must manually push a *reset* button.

Occasionally *transients* (brief, irregular signals) can cause a fuse to blow, or a circuit breaker to open, even if there is no defect in the load circuit at all. But if a new fuse immediately blows, or if a circuit breaker repeatedly re-opens right after it's reset, it indicates that something definitely is wrong with the circuit, and repairs are needed. **Never replace a fuse with a higher-rated unit.** You could easily end up blowing out some expensive electrical components or start a fire.

Fuses generally aren't necessary in battery-powered circuits, but anytime you are running a circuit off of ac house current, a fuse is a good idea, whether it is shown in the published schematic or not.

2
Transistor Switching

IN MANY PRACTICAL APPLICATIONS, AN ELECTRONIC SWITCHING CIRCUIT WILL BE highly desirable over a mechanical switching system. Either an electronic switching circuit or a mechanical relay circuit, can be automated. Electronic switching circuits tend to be smaller and lighter than their comparable mechanical counterparts. They also can operate faster and more reliably because there are no moving parts to wear out or get stuck in an electronic switch.

A bipolar transistor is designed primarily as a linear amplifier, but it also can be set up to operate as an electronic switch. There are several different types of transistor switching circuits. This chapter and the next will cover the basic principles of transistor switching.

THE THREE BASIC SWITCHING MODES

To get a transistor to function as an electronic switch, it must first be *biased* correctly. That is, the proper operating voltages must be applied (in the correct polarities) to the transistors three terminals.

There are three basic biasing schemes, or *modes* for transistor switching circuits. They are called:

> Saturated mode
> Current mode
> Avalanche mode

Each of these transistor switching modes will be described and discussed in the following pages.

Saturated Mode Transistor Switches

In the saturated mode, the transistor is turned on by biasing it into full current conduction, or *saturation*. The collector current in a saturated mode circuit is limited only by the resistances in the collector and emitter circuits. The voltage across the transistor (known as the *saturation voltage*) is at a minimum. The exact level of the saturation voltage is defined by the circuit's collector current and the load resistance.

The off condition in a saturation mode switching circuit is achieved by biasing the transistor so that no collector current flows. In this type of circuit, the collector current is either at its maximum value (saturation), or at zero.

A simple saturation mode transistor switching circuit is illustrated in FIG. 2-1. Battery (or other voltage source) V_{BB} biases the transistor into cutoff when no input signal is present. Under these conditions, the transistor's base is made negative with respect to its emitter. If a sufficiently positive voltage is applied to the input of this circuit, it will overcome this negative bias voltage, switching the transistor on and permitting the collector current to flow.

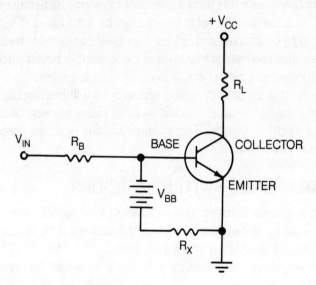

Fig. 2-1. A simple saturated mode transistor switch.

A voltage will be dropped across the load resistance R_L in this circuit only when there is some collector current. Since the transistor is biased so that there is a collector current only when the transistor is turned on (by a positive voltage at the input), there will be a voltage drop across R_L only when the transistor is in its on condition. Otherwise, the output voltage across the load resistance will be zero.

Ideally, the transistor should switch on and off instantly, with no transition time between the two states. This ideal simply isn't possible. Some finite time always will be required for the transistor to reverse its output state. In most applications this transition time will be so short that it will be negligible, but in some high-speed systems, the transition time may be of significance.

In the following discussion, assume that you are working with the circuit shown in FIG. 2-1. Initially, the transistor is in its off condition. There is no collector current, so there is no voltage across the load (R_L).

At some specific time (T_O) a positive pulse is applied to the input of the circuit. A base current will start to flow through the transistor right away, but there will be a brief period of time before the transistor's emitter/base voltage can climb from its initial negative value to a positive voltage. Collector current can not begin to flow until the emitter/base voltage is slightly positive. Once it starts to flow, the collector current will require some small period of time to reach its maximum level.

The *turn on time* is generally considered to be the total period of time (T_O) from when the positive pulse first is applied to the base, to when the collector current reaches 90% of its maximum value (I_L). This period is sometimes called the *delay time* of the circuit.

Similarly, a practical transistor cannot fall instantly from saturation down to cutoff when the input voltage is removed. When the positive input voltage is removed from the circuit, the bias voltage (V_{BB}) brings the voltage level on the transistor's base back down to a negative level. The base current goes negative until the emitter/base voltage goes negative and the emitter/base junction ceases to conduct. The emitter/base voltage and the collector current remain positive for a brief period of time after the base has been brought down to a negative voltage.

The finite time period required for the collector current to drop to 10% of its maximum level after the positive input signal is removed from the circuit is called the *storage time*. The storage time is determined primarily by the internal capacitances in the transistor. These internal capacitances are charged up when the transistor is turned on, and discharge when the transistor is turned off. Because these internal capacitances are extremely small, delays are in the microsecond (millionths of a second) or millisecond (thousandths of a second) range.

Bridging a capacitor C_B across the base resistor R_B, as shown in FIG. 2-2, increases the transistor's switching speed. This capacitor is not absolutely essential for the operation of the circuit, and it can be omitted in some less critical applications.

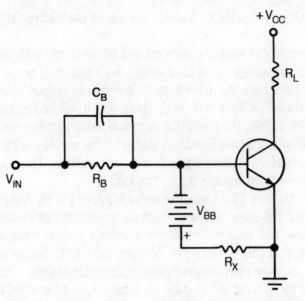

Fig. 2-2. A capacitor bridged across the base resistor can increase the transistor's switching speed.

To find the desired component values in this circuit, you must first know the saturation current, usually identified as $I_{C(SAT)}$. This value can be easily found via Ohm's Law:

$$I_{C(SAT)} = V_{CC}/R_L$$

where V_{CC} is the power supply voltage for the circuit and R_L is the output load being driven by the switching circuit.

In our design example, assume the following factors:

V_{CC} = 9 volts
V_{BB} = 1.5 volt
R_L = 10KΩ(10,000 ohms)

In this case the saturation current works out to a value of:

$$\begin{aligned} I_{C(SAT)} &= 9V/10000Ω \\ &= 0.0009 \text{ ampere} \\ &= 0.9 \text{ mA} \end{aligned}$$

This is a fairly typical value for the saturation current in a circuit of this type.

The necessary value of resistor Rx is found using this form of Ohm's Law:

$$R_X = V_{BB}/I_{CBO}$$

where V_{BB} is the negative base bias voltage, and I_{CBO} is the collector-to-base leakage current when the transistor is operating at its maximum rated temperature. This value can be obtained from the manufacturer's specification sheet for the transistor being used in the circuit. In this design example, assume that I_{CBO} has a value of $2\mu A$ (0.000002 amp), then Rx should have a value of:

$$R_X = 1.5V/0.000002A$$
$$= 750000 \text{ ohms}$$
$$= 750K\Omega$$

In the next step of the design process, you need to look at a characteristic curve graph for the transistor. A characteristic curve graph illustrates how the collector current (I_C) varies with the collector/emitter voltage (V_{CE}) for different values of base current (I_B). Characteristic curve graphs usually are included on the manufacturer's specification sheet. A typical characteristic curve graph for the transistor used in the design example is shown in FIG. 2-3.

Plot the load line for the collector circuit (the solid line marked *LOAD* in the graph). An estimated value for the base current (I_{BS}) is indicated by the point where the load line crosses the transistor's saturation resistance curve. In this example, I_{BS} is just a little greater than I_{B4}. You now can find the value you need for the base resistor (R_B), again using Ohm's Law:

$$R_B = V_{IM} / (I_{BS} - I_{CBO})$$

where V_{IM} is the maximum input voltage.

For the design example, assume that I_{BS} works out to 0.0225 mA (0.000225 amp) and the maximum input voltage (V_{IM}) is 2.50 volts. You already have determined that the collector-to-base leakage current (I_{CBO}) is $2\mu A$ (0.000002 amp). In these circumstances, the desired value for the base resistor (Rb) works out to:

$$R_B = 2.5V/ (0.000225A - 0.000002A)$$
$$= 2.5V/0.000223A$$
$$= 11211 \text{ ohms}$$
$$\cong 12000 \text{ ohms}$$
$$= 12K\Omega$$

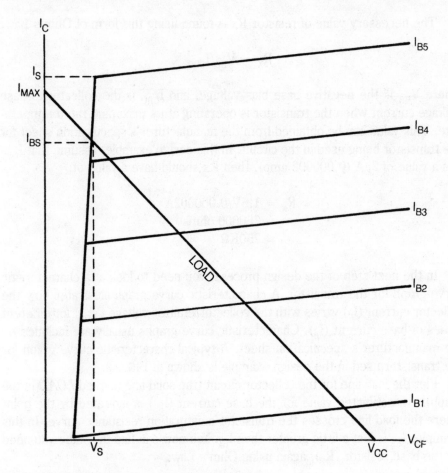

Fig. 2-3. A typical characteristic curve graph for a transistor.

Notice that the value of R_B is rounded off to the nearest standard resistance value (12KΩ). The exact values of the resistors in this type of circuit usually aren't critical, so such rounding off is almost always permissible.

If the base capacitor (C_B) is included in the circuit, its value could be derived mathematically, but it is usually easier just to breadboard the circuit and experiment with different capacitance values until a satisfactory switching speed is obtained. After you have worked with switching circuits like this a few times, you will get a pretty good feel for good capacitor values to try.

Current Mode Transistor Switches

The second basic type of transistor switching circuit is the current mode switch. Higher switching speeds can be obtained if the transistor is not put into saturation when it is turned on. In the current mode, the transistor is biased so that it operates close to, but not quite at, the saturation point. The collector-emitter voltage is therefore somewhat less than the saturation voltage of the device.

As in the saturated mode circuit discussed previously, the current mode transistor switch is turned off by biasing it so that the transistor is in the cutoff (nonconducting) state. A typical current mode switching circuit is shown in FIG. 2-4. Notice how similar this circuit is to the saturated mode switching circuit shown in FIG. 2-1.

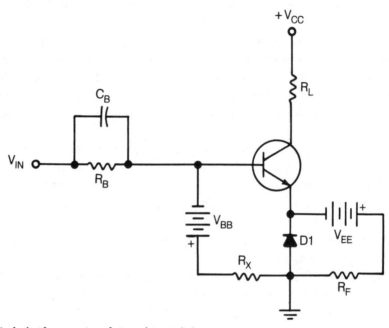

Fig. 2-4. A simple current mode transistor switch.

The battery (V_{BB}) and resistor (R_x) combination in the base circuit holds the transistor in cutoff when there is no input signal. This works in essentially the same way as in the saturated mode circuit described earlier. The transistor is held in cutoff by applying a negative bias voltage on the base.

The emitter voltage source (V_{EE}) keeps diode D1 turned on (forward-biased and conducting) at all times. Assuming a silicon diode (the most common type) is used, there will be a voltage drop of approximately 0.7 volt across this component. If the base of the transistor is at ground potential (0 volts between the base and ground), only the 0.7 volt across the diode will be between the emitter and the base. This small voltage would be sufficient to keep the transistor turned on (conducting).

To turn on the current mode transistor switch, the base bias voltage (V_{BB}) has to be canceled out by an input signal with the opposite polarity. The input signal must have an absolute value equal to or greater than the bias voltage (V_{BB}). Whenever the input voltage (V_{IN}) is equal to or greater than V_{BB} (opposite polarities are assumed), the transistor is switched on. When V_{IN} is less than V_{BB}, the transistor is turned off.

If the emitter voltage source (V_{EE}) was not included in the circuit, we would have virtually the same situation as in the saturated mode switching circuit of FIG. 2-1. When the transistor is turned on by a positive pulse at the input, the collector current will be equal to the beta of the transistor multiplied by the base current. If the base current is large enough, the transistor will be driven into saturation.

The three added components in the emitter circuit (V_{EE}, R_E, and D_1) limit the input current in the circuit of FIG. 2-4. Because the input current is limited, the collector current is limited. The maximum current through the transistor in this circuit is less than its saturation value. The maximum current flow through this circuit's transistor is defined by the values of the three added components in the emitter circuit.

The diode's polarity prevents any current from the emitter from flowing through it. But the 0.7-volt drop across the diode does limit the current flow through the emitter resistor (R_E). The current level flowing through this resistor is the maximum amount of current that can flow through either the emitter or the collector in this circuit. The transistor will not be driven into saturation, provided that the saturation current value ($I_{C(SAT)} = V_{CC}/R_L$) is greater than the current flow through R_E (V_{EE}/R_E).

In most practical circuits, the current through the load resistor (R_L) should be limited to a maximum level equal to or less than the limited current flowing through the emitter resistor (R_E). In designing the sample saturated mode switching circuit in the preceding section, you found that the saturated collector current (the transistor's on current) was 0.9 mA. For a current mode switching circuit, you obviously want the emitter limiting current ($I_{EL} = V_{EE}/R_E$) to be less than this saturation value.

In the following example, assume that you want a current limit of 0.7 mA (0.0007 amp). If V_{EE} is 1.5 volts (a typical value for this type of circuit), the emitter resistor (R_E) should have a value of about:

$$R_E = V_{EE}/I_{EL}$$
$$= 1.5V/0.0007A$$
$$= 2143 \text{ ohms}$$

This is not a standard resistance value, but you can round off. If you use a standard 2.2K (2200 ohms) resistor for R_E, the maximum collector/emitter current (ignoring component tolerances) will work out to approximately:

$$I_{CMAX} = V_E / R_E$$
$$= 1.5V/2200\Omega$$
$$= 0.00068 \text{ amp}$$
$$= 0.68 \text{ mA}$$

As you can see, rounding off the value of R_E did not alter the desired limited current value appreciably.

Since the collector current is limited to 0.68 mA, the transistor will not be put into saturation when it is turned on. Because the transistor does not have to work quite so hard (conduct as much current) in this circuit, it can switch between states considerably faster than the same transistor could in a comparable saturated mode switching circuit. The reduced current levels flowing through the transistor also will minimize heat buildup and reduce the odds of any thermally related problems showing up in the operation of the circuit.

Avalanche Mode Transistor Switches

The third and final transistor switching mode is the *avalanche mode*. This is the fastest of the three operating modes for a standard bipolar transistor. Even faster switching can be obtained with specialized semiconductor devices, such as hot-carrier, pin, snap-off, or tunnel diodes, but they are beyond the scope of this book.

In the avalanche mode, the on and off states of the transistor switch are kept within the breakdown portion of the transistor's operating curve. Saturated mode switching circuits and current mode switching circuits both require a specific base voltage (bias) to be maintained to hold the transistor off in the absence of a triggering input signal. The trigger signal must be present continuously to hold the transistor switch in its on state. In an avalanche mode switching circuit, however, a brief trigger pulse is all that is needed to put the transistor in either its on or its off state. The transistor switches into the desired input state almost instantly upon receiving the necessary input trigger pulse.

In terms of component layout, an avalanche mode switching circuit is very similar to a saturated mode circuit. A typical avalanche mode circuit is illustrated in FIG. 2-5. Notice that, unlike the saturated mode circuit discussed earlier in this chapter, this circuit doesn't need a base capacitor (C_B) to improve the switching speed. The avalanche mode switching circuit is very fast, even without a base capacitor.

One of the differences between the avalanche mode circuit and the saturated mode circuit is that the V_{CC} voltage is much higher for the avalanche mode cir-

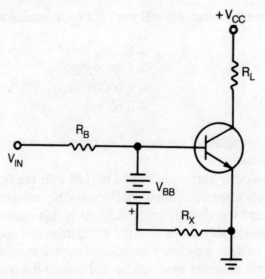

Fig. 2-5. A simple avalanche mode transistor switching circuit.

cuit. This forces the transistor to operate in the voltage breakdown region of its collector characteristic, as illustrated in the curve graph of FIG. 2-6. Three collector current curves are shown in this graph. Each curve represents a different amount of base current:

Curve	Bias Current
1	0 mA (V_{BB} + positive input)
2	slightly negative (V_{BB})
3	very negative (V_{BB} + negative input)

For curve 3, it is assumed that the base current is not negative enough to actually damage the transistor. The unlabeled curves in FIG. 2-6 represent the collector characteristic when the base current is positive.

Normally, the transistor is held in an *idling* state by the negative bias voltage applied to its base (V_{BB}). Assume that the transistor is biased so it idles at point A on curve 2. If an external positive voltage is applied to the base, overcoming the negative bias voltage, the transistor's idling point will be pushed up to a new curve (to the left on the graph). The exact location of this second curve will depend on the relative magnitudes of the positive input voltage and the negative bias voltage.

If you assume that the positive input voltage is of the exact same magnitude (but the opposite polarity) as the negative bias voltage, then the effective base voltage now will be zero. This will place the shifted idling point on curve 1 at the

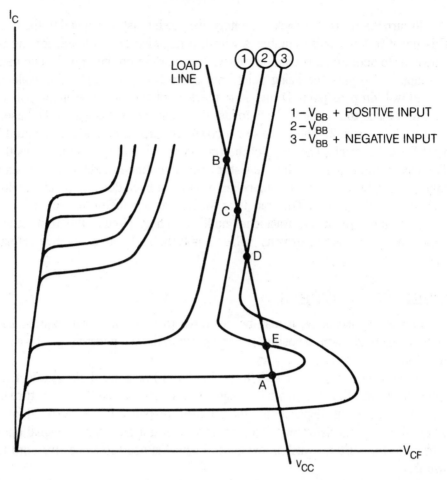

Fig. 2-6. In the avalanche mode, the transistor is forced to operate in the voltage breakdown region of its collector characteristic.

point marked B in the graph of FIG. 2-6. This level will be held as long as the positive input signal is maintained on the base. When the positive input signal is removed, the idling point will shift to point C, which is on curve 2. It reverted to curve 2 because only the negative bias voltage was being applied to the base.

The interesting feature here is that the transistor will continue idling at point C even without any additional positive input signal. This state is stable because the slope of the load line is not as steep as the slope of the transistor curve. In most circuits, a point on this part of the curve would not be stable because the transistor exhibits negative resistance, but in this type of circuit, the relative slopes of the load line and the transistor curve change the situation and make this point quite stable.

To turn the transistor back off, a negative pulse must be applied to the base. This negative input pulse combines with the negative bias voltage, forcing the transistor to operate on yet another curve (to the right on the graph). The negative input pulse puts the idling point onto curve 3 of the graph. The collector current will drop to point D and stay there until the negative input pulse is removed. Since the voltage is back to just the original bias voltage on the base, it goes back to curve 2. The transistor's collector current is now represented by point E on the graph. The slope of the transistor curve at this point is less than the slope of the load line. The transistor cannot continue to idle at this unstable point. It quickly will drop its idling collector current back down to point A, which is right where it started. The transistor switch has now been turned off.

To summarize, in an avalanche mode switching circuit, the transistor is turned on with a positive external pulse. It is turned off with a negative external pulse.

INVERTING SWITCHES

In some applications, it may be desirable to invert the switching polarity. That is, you might want a positive control pulse to drive a negative-going switching action.

In most basic transistor switching circuits, a low control signal effectively opens (turns off) the switch, while a high control signal is used to close (turn on) the switch. In certain applications, the opposite response might be called for; in other words, a transistor switching circuit in which a high control signal opens (turns off) the switch, which a low control signal is used to close (turn on) the switch.

Fortunately, this is very easy to accomplish with transistor switching circuits. A very simple inverting transistor switching circuit is illustrated in FIG. 2-7. As you can see, the circuitry is very simple, and perfectly straightforward.

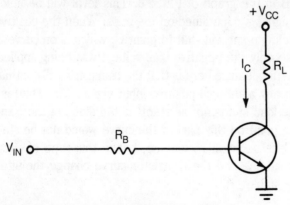

Fig. 2-7. This transistor switching circuit inverts the signal polarity.

Typical input and output signals for this circuit are shown in FIG. 2-8. This figure clearly demonstrates that the circuit does precisely what we want it to do—it inverts the switching action. A positive voltage on the transistor's base causes both the collector current and the collector voltage to drop to zero. A positive control signal at the base turns the transistor off.

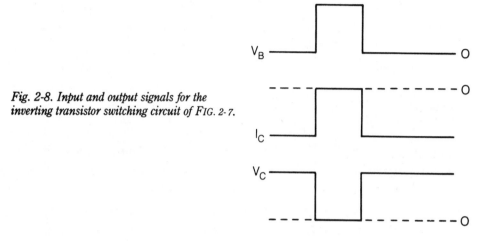

Fig. 2-8. Input and output signals for the inverting transistor switching circuit of FIG. 2-7.

SELECTING TRANSISTORS FOR SWITCHING CIRCUITS

In most applications, almost any transistor can be used in a switching circuit, provided it can handle the necessary voltage and current levels demanded by the intended load. A cheap transistor from a grab bag is likely to work every bit as well as an expensive transistor.

Unlike amplifier applications, many transistor specifications really don't matter at all in a switching circuit. The frequency response and linearity of the transistor are virtually irrelevant. The one exception is in certain high-speed switching circuits. Obviously, the frequency response of the transistor must be sufficient to accept and respond to the intended switching rate.

The only other thing of concern in choosing a transistor for a switching circuit is that the circuit provides a sufficient control signal to turn the transistor on and off reliably. If possible, use the characteristic curves for the specific device used from the manufacturer's specification sheet. For most hobbyist applications, you can breadboard the circuit and experiment with the control signal levels until the circuit operates as desired.

3

Two Transistor
Switches

FOR MANY PRACTICAL SWITCHING APPLICATIONS, THE SIMPLE SINGLE TRANSIS-
tor circuits presented in chapter 2 will do the job just fine. In some cases, especially where large loads are involved, a single transistor simply can't handle sufficient power to drive the desired load properly. In such heavy-current switching
applications, it is often necessary to use more sophisticated circuits, using two
(or sometimes more) transistors. Several different two transistor switching circuits will be discussed in this chapter.

LARGE LOAD SWITCHING

A two-transistor switching circuit typically is used when either the current
or the voltage to be switched is too large to be handled by a single transistor.
Two-transistor switching circuits also permit more convenient and reliable polarity inversion (see chapter 2).

A very simple, but practical two-transistor switching circuit is illustrated in
FIG. 3-1. Typical operating signals for this type of circuit are shown in FIG. 3-2. In
this circuit, the polarity of the load current (I_{C2}) is inverted, but the load voltage
(V_{C2}) has the same polarity as the input signal.

The first transistor (Q1) in this circuit does the actual switching, while the
second transistor (Q2) serves as an amplifier, or *buffer* for the output signal of
the first transistor. When a suitable positive voltage is applied to the circuit input
(base of transistor Q1), the first transistor (Q1) is turned on, and the second
transistor (Q2) is turned off.

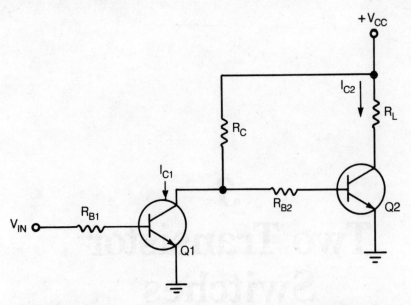

Fig. 3-1. Two transistors are sometimes used in switching circuits to handle larger loads.

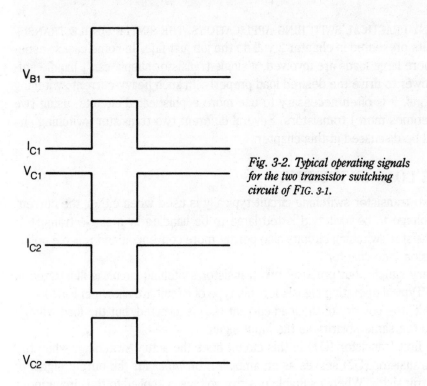

Fig. 3-2. Typical operating signals
for the two transistor switching
circuit of FIG. 3-1.

Q1's collector is brought low when it is turned on. Current flows through resistor R_C, but not through resistor R_{B2}, so transistor Q2 is cut off. Its collector voltage goes high, and the collector current from this transistor is shut down. Virtually no current can reach the load (R_L) under these conditions.

The best approach to designing this type of circuit for a specific application is to start at the load (and its operating requirements) and work backwards through the circuit to the input. This is because we can modify the available input signal to suit the circuit easily enough (with an amplifier stage, or an attenuating voltage divider), but the intended load demands the correct current and voltage be available when it is switched on.

Transistor Q2 must be selected to safely handle the power levels (both current and voltage) required by the circuit's intended load. A heavy-duty power transistor often will be used here. Once we've selected a Q2 transistor that can handle sufficient power, we can just look at the manufacturer's specification sheet and determine the required base current (see chapter 2). Transistor Q1 must put out sufficient current to drive the base of transistor Q2. Now we know how much power the first transistor in the circuit must handle.

There is one potentially confusing feature here. The base resistor for transistor Q2 actually is made up of two real resistors, Rc and R_{B2}. The effective base resistance for Q2 is the series combination of these two resistors. In some applications, resistor R_{B2} may be omitted from the circuit. In this case, only transistor Q1 is supplying current to transistor Q2. The second transistor is not directly drawing any current from the circuit's power supply (V_{CC}).

The value of resistor R_C is rather important in this circuit. This resistance controls the level of current flowing through Q1's collector when Q1 is on. For the best results, do not allow too much current to flow here. In the vast majority of applications, a few milliamps will do the job just fine. Only extremely high power circuits will ever require a higher collector current from Q1. Keep the current requirements for this transistor small, and let Q2 do the heavy work. This will keep the overall circuit size and cost down. Transistor Q1 does not have to put out much current to reliably control transistor Q2.

A slightly different approach to a two-transistor switching circuit is shown in FIG. 3-3. Typical input and output signals for this circuit are illustrated in FIG. 3-4.

When a positive input pulse is received, the first transistor (Q1) is switched off. This brings the base of the second transistor (Q2) down to a low voltage, turning this device on. The collector voltage goes high and collector current flows through the load. The input pulse must be close to the circuit's supply voltage to reliably and completely turn off transistor Q1.

A nice, though unobvious, feature of this switching circuit is that the load is isolated from the power supply line. This means that multiple switching circuits can be used together in a hard-wired OR configuration, as illustrated in FIG. 3-5.

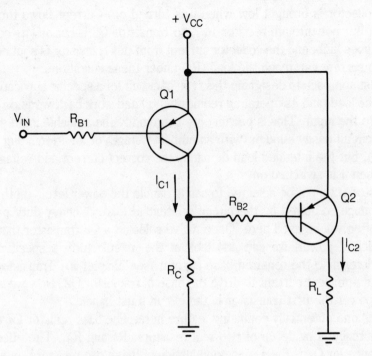

Fig. 3-3. An alternate two transistor switching circuit.

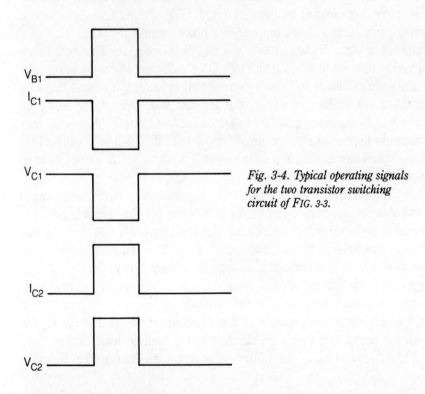

Fig. 3-4. Typical operating signals
for the two transistor switching
circuit of FIG. 3-3.

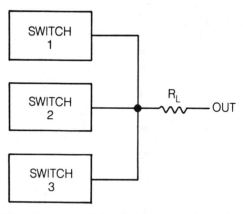

Fig. 3-5. Multiple switching circuits can be hard-wired together in an OR configuration.

Current is supplied to the load whenever any one of the switching circuits is activated. The hard-wired OR configuration is very easy to set up. Simply tie all of the output transistors' (Q2) collectors together across the load device or circuit.

There is one important restriction involved in using the hard-wired OR configuration for a circuit of this type. The system should be designed so that only one of the switching circuits is activated at any given time. In some applications, if more than one of the switching circuits is in the on mode, the load could be damaged.

COMPLEMENTARY PAIR SWITCHING CIRCUITS

The two-transistor switching circuit of FIG. 3-1 used two NPN transistors, while a pair of PNP transistors were employed in the switching circuit of FIG. 3-3. In some applications it may be highly desirable to mix transistor types, as in the switching circuit of FIG. 3-6. In this circuit there is one NPN transistor and one PNP transistor.

Using complementary transistor types permits the circuit to operate with no polarity inversion of the signal. The controlling pulse has the same polarity as the output switching action. This type of circuit is also highly sensitive. Typical input and output signals for the complementary pair transistor switching circuit of FIG. 3-6 are shown in FIG. 3-7.

A positive input pulse turns transistor Q1 on. The collector voltage and current of this device go low, effectively grounding the base of transistor Q2. This condition forces the second transistor (Q2) to turn on also. Current flows through the collector of Q2, and through the load.

This complementary transistor switching circuit also isolates its load from the circuit's supply voltage. Multiple circuits can be used together in a hard-wired OR configuration. Current is supplied to the load device, or circuit, when-

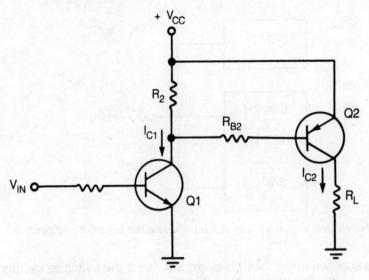

Fig. 3-6. In some switching circuits, NPN and PNP transistors are used together.

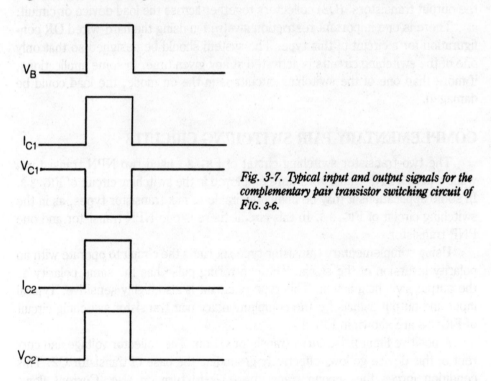

Fig. 3-7. Typical input and output signals for the complementary pair transistor switching circuit of FIG. 3-6.

ever any one of the switching circuits is activated. The hard-wired OR configuration is very easy to set up. Simply tie all of the output transistors' (Q2) collectors together across the load.

The hard-wired OR system should be designed so that only one of the switching circuits is activated at any given time. In some applications if more than one of the switching circuits is in the on mode, the load could be damaged.

Complementary transistors also make it convenient to electronically switch negative voltages. A typical switching circuit of this type appears in FIG. 3-8. When transistor Q1 is in its off condition, its collector is high, holding the second transistor (Q2) also in cutoff. The base of Q2 must go more negative (by at least 0.7 volt) than its emitter in order to turn this transistor on.

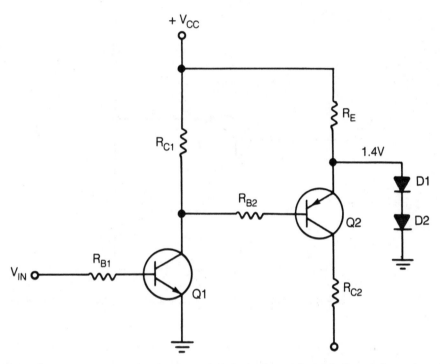

Fig. 3-8. Complementary transistors also make it convenient to electronically switch negative voltages.

The Q2 emitter voltage is clamped at about +1.4 volts by the two diodes (D1 and D2). This means that Q2's base must be fed a signal no greater than +0.7 volts for this device to be switched on. This switching circuit is very sensitive. The diodes are used in this circuit to ensure against accidental turn-on. If the diodes were not clamping the Q2 emitter voltage at +1.4 volts, leakage through transistor Q1 could turn on transistor Q2 when we want it to remain off. To use this switching circuit with a very heavy load, resistor R_{C2} could be replaced with a third transistor circuit.

DARLINGTON TRANSISTORS

Bipolar transistors often can become quite unstable if a high-output current is required of them. A more stable high-gain circuit can be created by connecting a pair of bipolar transistors in series, as shown in FIG. 3-9. The emitter of the first transistor (Q1) feeds the base of the second transistor (Q2). When two bipolar transistors are interconnected in this fashion, they are called a *Darlington pair*.

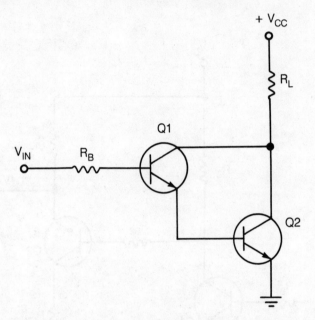

Fig. 3-9. A more stable high-gain circuit created by connecting a pair of bipolar transistors in series.

For the best performance in a circuit, the two transistors used in a Darlington pair should be very closely matched. Obviously, they should be of the same type number. In many circuit applications a Darlington pair can be used almost as if they were a single "super" transistor. The current at the emitter is virtually the same as the current at the common collectors. This allows for excellent balance.

Because the Darlington pair is so convenient in many practical circuits, a number of manufacturers produce specialized *Darlington transistors*. A Darlington transistor is essentially an extremely closely matched pair of individual transistors contained within a single housing. Externally, a Darlington transistor looks like a regular transistor. In some cases, it may be slightly larger than an ordinary bipolar transistor of similar power ratings.

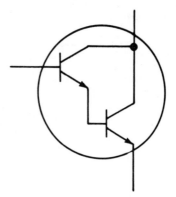

Fig. 3-10. A Darlington transistor is a pair of series-connected bipolar transistors in a single housing.

The schematic symbol for a Darlington transistor is shown in FIG. 3-10. To indicate that this is a single unit, rather than two discrete (separate) transistors, the circle around the symbol almost always is included. For other transistor types, the circle in the schematic generally is considered optional.

Darlington transistors can be used to switch moderately large loads, using a circuit along the lines illustrated in FIG. 3-11. In designing such a circuit, the Darlington transistor should be selected to handle the desired voltage and current required by the intended load device or circuit. The value of the base resistor (R_B) is calculated in the same manner as for the simpler one transistor switching circuits discussed in chapter 2.

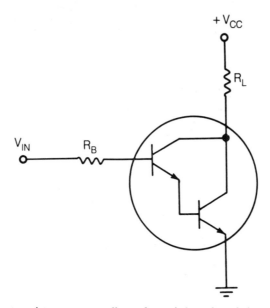

Fig. 3-11. Darlington transistors are generally used to switch moderately large loads.

4
FET Switches

IN CHAPTERS 2 AND 3 SWITCHING CIRCUITS WERE BUILT AROUND BIPOLAR TRAN-
sistors. But there are other types of transistors too, and these devices can also
be used in electronic switching circuits.

More and more circuits these days are replacing bipolar transistors with
FETs, which have operating characteristics more similar to those of old-
fashioned vacuum tubes.

WHAT IS A FET?

The term FET is an acronym, standing for *Field Effect Transistor*. In its
structure, a bipolar transistor is basically a semiconductor sandwich, as illustrated in FIG. 4-1. Compare this with the basic structure of a FET, shown in FIG.
4-2. The N-type FET's body is a single continuous length of N-type semiconductor material. One end of this semiconductor slab is brought out to a lead called
the *source*. The opposite end terminal is known as the *drain*. A small section of
P-type material is placed on either side of the larger N-type section. Both of
these P-type sections are tied electrically together within the device. The connection to the P-type material is called the *gate*.

P-type FETs are also available, although they are relatively rare. In this type
of device, the large central section is made up of P-type material, and the small
sections on the sides are made up of N-type semiconductor material. All operating polarities are reversed for a P-type FET.

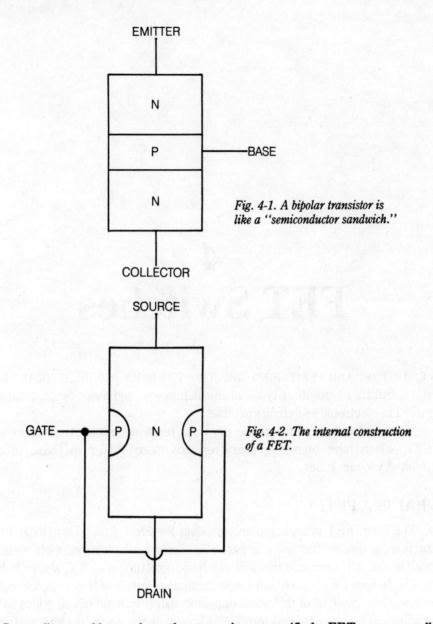

EMITTER

N

P — BASE

N

Fig. 4-1. A bipolar transistor is
like a "semiconductor sandwich."

COLLECTOR

SOURCE

GATE — P   N   P

Fig. 4-2. The internal construction
of a FET.

DRAIN

Generally speaking, when the type is unspecified, FETs are usually assumed to be of the N-type. The standard schematic symbol for a (N-type) FET is shown in FIG. 4-3. In some schematic diagrams, the circle around the symbol is omitted. The FET has three leads, labeled as follows:

S   Source
G   Gate
D   Drain

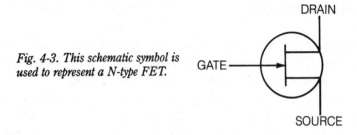

Fig. 4-3. This schematic symbol is
used to represent a N-type FET.

The symbol for a P-type FET appears in FIG. 4-4. Notice that the only difference is the direction of the arrow. The same three lead labels (Source, Gate, and
Drain) are used.

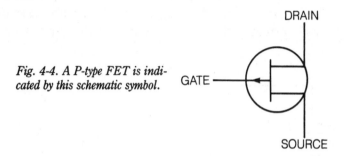

Fig. 4-4. A P-type FET is indicated by this schematic symbol.

To get a general idea of how a FET works, compare it to the mechanical
water flow system illustrated in FIG. 4-5. When the valve (or gate) is opened, as
in FIG. 4-5A, water can flow through the pipe (from the source to the drain.) If, on
the other hand, the valve (gate) is partially closed, as in FIG. 4-5B, the amount of

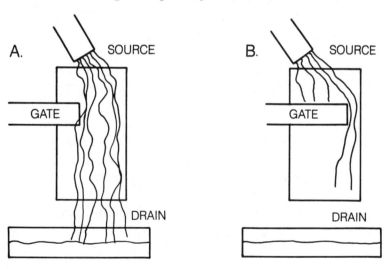

Fig. 4-5. A FET is conceptually similar to a mechanical water flow system.

water that can flow through the pipe is limited. Less water comes out of the drain.

A FET is electrically similar to this system. Instead of water, a FET controls electrical current. The signal on the gate terminal of a FET controls the amount of electric current that can flow from the source to the drain.

Like all transistors, a FET must be biased properly to function correctly. The correct biasing for a N-type FET is illustrated in FIG. 4-6. A negative voltage applied to the gate terminal reverse biases the PN junction, producing an *eletrostatic field* (or electrically charged region) within the N-type material of the FET. This electrostatic field opposes the flow of electrons through the N-type section, acting somewhat like the partially closed valve in the mechanical model (FIG. 4-5). The higher the negative voltage applied to the gate, the less the amount of current that is allowed to pass through the device from the source terminal to the drain terminal. The current path from the source to the drain sometimes is called the *channel*.

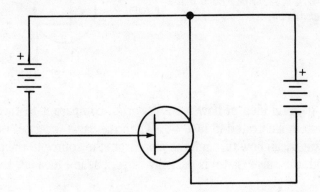

Fig. 4-6. A FET must be correctly biased to function properly.

The operation of a FET is very similar to the action of a vacuum tube. The three leads of this semiconductor device correspond directly to the three leads of a standard triode tube. The gate of the FET corresponds to the tube's grid, controlling the amount of current flow through the device. The source in a FET is the equivalent to the cathode of a tube. The source and the cathode each act as the source of the electron stream through the device. Finally, the drain of a FET serves the same functions as a tube's plate, draining off electrons from the device.

FETs have a very high input impedance, especially as compared to ordinary bipolar transistors. Thanks to this high-input impedance, a FET draws very little current from the driving circuit or device. This is in accordance to Ohm's Law:

$$I = E/Z$$

where I is the current, E is the voltage, and Z is the impedance. Using this equation, you can see that increasing the impedance while holding the voltage constant will decrease the current.

The high-input impedance of a FET means that it can be used in highly sensitive measuring applications. They are good in any circuit where it is important to avoid loading down (ie, drawing heavy currents from) previous circuit stages. This is another way in which FETs more closely resemble vaccuum tubes than bipolar transistors.

FET SWITCHING CIRCUITS

A FET can be used in switching applications. This type of component offers both advantages and disadvantages compared to bipolar transistors in such applications. The biggest disadvantage of a FET in a switching circuit is a lower limit on the maximum switching speed. Relatively large internal capacitances within a FET restrict the device's high frequency response. In many practical switching applications, you really won't need a switching speed faster than a FET can handle easily, so this restriction usually isn't a problem.

A very simple FET switching circuit is illustrated in FIG. 4-7. Compare this circuit to the bipolar transistor switching circuits presented in chapter 2. FIGURE

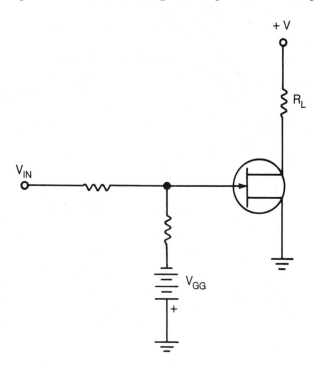

Fig. 4-7. FETs can be used in switching circuits.

4-8 shows a typical characteristic curve graph for a FET. Refer to this graph for the discussion of the operation of the switching circuit shown in FIG. 4-7.

Initially, assume that the input signal to this circuit (V_{IN}) is zero. (This voltage may be slightly negative without changing anything.) The V_{GG} voltage supply biases the FET with a negative voltage between the source and the gate. If this negative bias voltage is more negative than the *pinch-off* voltage for the FET being used in the circuit (refer to the manufacturer's specification sheet for the particular device), the FET will be essentially cut off. Very little current will be able to flow through the drain circuit. The load (R_L) will not see any signal from the FET. The drain current under these circumstances will be negligible.

Now consider what happens if a positive voltage is fed into the circuit as V_{IN}. This input signal changes the V_{GS} (gate-to-source) voltage seen by the FET. A positive input signal to this circuit puts the operation of the FET at the high end of the I_D (drain current) range. This can be seen easily in the characteristic curves of FIG. 4-8. Under these circumstances, the drain circuit's conduction is at its maximum level. Current flows through the load (R_L).

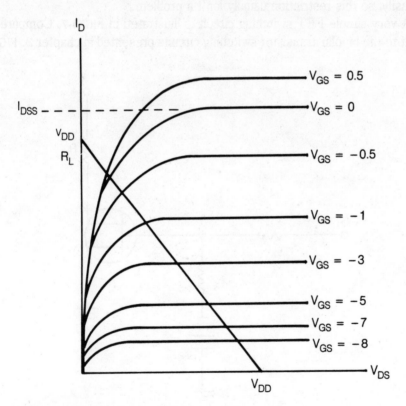

Fig. 4-8. A typical characteristic curve graph for a FET.

A FET switching circuit alternately can be thought of in terms of resistances. When the circuit is in its off mode (the FET is held at pinch-off), the effective resistance is extremely high. On the other hand, when the device is switched on, the drain resistance drops to a very low value. It is often useful for a circuit designer to look at the FET switch in this manner to determine how best to drive certain loads.

5
SCR Switches

A SPECIAL CLASS OF SEMICONDUCTOR COMPONENTS KNOWN AS *THRYISTORS* are used primarily in switching applications. The most important and commonly used of the thyristors is the *SCR*, or *Silicon Controlled Rectifier*. The SCR is a three-lead device, like the common bipolar transistor.

WHAT IS AN SCR?

The SCR is easy to understand, just look at each individual word in its name. *Silicon* is the semiconductor material the SCR's active portion is made of. The third word in the term, *Rectifier*, is also perfectly straightforward. It indicates that the device is a diodelike component, capable of rectification. The important term here is in the one in the middle—*controlled*. The rectification of an SCR is controlled externally via the added third terminal, which is called the *gate*. The other two terminals carry over the names *anode* and *cathode* from regular, two-terminal diodes. You could consider an SCR to be an electrically switchable diode. The SCR generally is used in ac or dc power control applications.

The most widely used standard schematic symbols for a SCR are illustrated in FIG. 5-1. The use of the surrounding circle is optional. Some technicians think the circle makes the symbol more clearly visible, while others feel it doesn't add any information to the diagram, and so is unnecessary. This is purely a matter of personal preference.

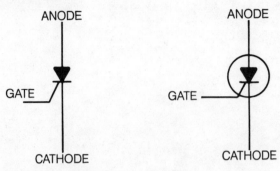

Fig. 5-1. The SCR is essentially an electrically switched diode.

HOW AN SCR FUNCTIONS

If a voltage is applied between the anode and the cathode of an SCR, as shown in FIG. 5-2, nothing will happen when there is no signal on the gate lead. In FIG. 5-2, the gate lead is shown grounded to emphasize that there is no signal being applied to this terminal. This is not a practical circuit. It is included here solely to demonstrate the operating concepts behind the SCR. In this circuit, the SCR acts like an open circuit.

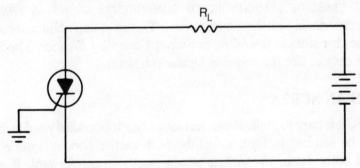

Fig. 5-2. In this circuit, the SCR looks like an open circuit.

But what happens if a voltage signal is applied to the gate of the SCR, as illustrated in FIG. 5-3? If the gate voltage is very low, there will be no change. The SCR continues to block current flow from its cathode to its anode.

Now assume that the gate voltage is being gradually increased. At some point it will exceed a specific level (determined by the internal construction of the SCR). At this point, the SCR will be triggered. The rectifier is activated. Current can now flow from cathode to anode against only a small internal resistance, just as with an ordinary diode. This current will continue to flow through the device, even if the voltage on the gate terminal is now removed. Once the SCR is switched on, it stays on, regardless of the gate signal.

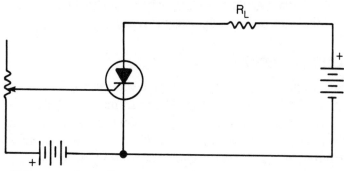

Fig. 5-3. A sufficient gate voltage will cause an SCR to start conducting.

The only way to stop the current flow through the SCR, once it has been started, is to decrease the positive voltage (with respect to the cathode) on the anode, or remove it altogether. When the anode voltage drops below a specific level (again, this value is determined by the SCR's internal construction), current flow will be blocked. The SCR turns itself off when it is reverse biased. Once it is off, it stays off, regardless of the voltage at the anode. Even if the anode is made positive again (with respect to the cathode), the SCR will remain off and will block current flow until it is again triggered by a suitable voltage at the gate.

The gate can turn the SCR on, but it cannot turn it off. Essentially, as stated earlier, an SCR is an electrically switched diode. This component is sometimes referred to as a *programmable diode*.

ON AND OFF STATES

The SCR has two stable states. These states usually are called *on* and *off*. In the on state, current can flow from the anode to the cathode, passing through a very low (almost negligible) resistance. In the off state, the current flow is blocked by a very high resistance between the anode and the cathode. Neither state normally permits current to flow from the cathode to the anode. The SCR is strictly a dc device, like any rectifier.

Both of these states are stable. Once the SCR is placed into one or the other of these states, it will hold that state until it is acted upon externally in very specific ways. In the on state, the SCR functions as a rectifier, permitting current flow in only one (forward-biased) direction.

To turn the SCR on, a voltage of a specific level must be fed to the gate lead of the device. When the voltage on this terminal exceeds the unit's trigger voltage, *breakover*, or switching will occur. Depending on the specific SCR being used, the maximum gate voltage typically is in the 2- to 5-volt range. It is very important to check the manufacturer's specification sheet before using any SCR or other semiconductor device, especially thyristors.

SCR SPECIFICATIONS

On some specification sheets, the trigger voltage may be labeled *breakover voltage* (V_{BO}) or *switching voltage* (V_S). These terms are interchangable. Don't be thrown by the change in terminology. In some areas of electronics there, unfortunately, isn't much standardization. Different technicians and manufacturers prefer different labels. Since the choice is arbitrary, it would make sense for everyone to agree on one fixed term for each parameter. This is unlikely to happen in the near future. Anyone working in electronics, either professionally, or as a hobby, needs to be aware of such inconsistencies in terminology.

In a practical circuit, the SCR should be rated for a maximum voltage higher than the peak voltage it ever will be exposed to in the circuit. The circuit designer should always leave plenty of extra elbow room. An excessive voltage fed through the SCR could damage or destroy it. This could also be harmful to any load circuit being driven by the SCR.

As a general rule of thumb, the SCR should be rated at least 10% to 25% higher than the absolute maximum anticipated voltage in the circuit. More, in this case, is always better. You can't use too high a voltage rating for a thyristor. The only trade offs are cost and the physical size of the component itself. There is also a very real risk of hazards to the circuit operator.

The maximum acceptable voltage that safely can be fed through a given SCR generally is labeled V_{DRM} on manufacturer's specifications sheets. This rating is more or less equivalent to the peak reverse voltage rating for a standard rectifier diode. The V_{DRM} rating usually will be at least 100 volts greater than the V_{BO} (or V_S) rating. In selecting an SCR for a specific application, leave plenty of headroom in the V_{DRM} rating. For example, if you are working with ac line current (120 volts ac), it would not be unreasonable to use an SCR rated at 600 volts.

During the turn-on process, a high instantaneous power level will be dissipated through the SCR. If this power level increases too much or too fast, the SCR could be damaged or destroyed. The manufacturer's specification sheet will include a dI/dT rating, sometimes written as $\Delta I / \Delta T$. The small triangle or the *d* in these terms represent the Greek letter *delta* which stands for a changing, rather than a static, value.

In working with SCRs (and other thyristors), you should be aware that junction leakage currents and current gain will increase with increases in temperature. The main effect of this is that the device can be triggered by a lower gate current. The gate itself may be regarded as a diode and will display a lessening voltage drop with increases in temperature.

TRIACS

SCRs, like diodes, are unidirectional. (You should be aware that this is not true for all types of thyristors.) SCRs have a definite polarity. Current can only flow in one direction. If the input is an ac waveform, the SCR can conduct for only half of each cycle, or less as illustrated in FIG. 5-4. For some applications, these characteristics are desirable. For other applications, they are a major disadvantage.

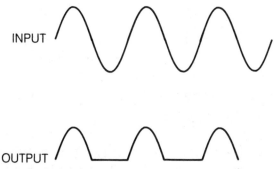

Fig. 5-4. An SCR conducts for only half of each ac cycle.

When a given application requires a bipolar current flow, an ordinary SCR just won't do the job. One possible solution would be to wire a pair of SCRs back to back, as illustrated in FIG. 5-5. This is effectively, a bidirectional SCR. Current can flow in either direction.

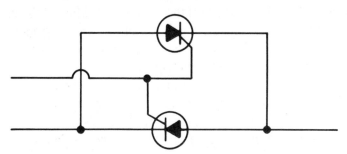

Fig. 5-5. Bidirectional current can be passed by a pair of back to back SCRs.

This trick is so useful, that semiconductor manufacturers make it unnecessary. They make self-contained specialized components which contain a pair of back to back SCRs in a single package. These devices are called *triacs*. The name is intimidating, but if you understand the SCR, you should have no problem comprehending and using the triac.

The schematic symbol for a triac is shown in FIG. 5-6. Notice that this symbol suggests back to back SCRs. Like an SCR, a triac has three leads. The gate lead serves the same function as on an SCR. It controls the current flow through the main body of the device. The signal applied to the gate terminal normally is used to turn the Triac on.

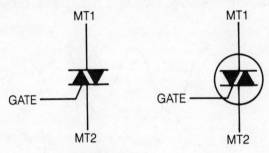

Fig. 5-6. The triac is essentially a bidirectional SCR.

Since current can flow between the other two leads in either direction, the terms anode and cathode are inappropriate. These leads on a triac usually are labelled *MT1* and *MT2*. These terms are abbreviations for Main Terminal 1 and Main Terminal 2. For the vast majority of practical applications, these two leads are fully interchangable, and the triac can be connected into a circuit in either direction.

When properly triggered, a triac can pass current in either direction. In a power controller circuit with an ac waveform as the input, a triac can pass anything from about 0% to almost 100% of the input signal. More of the input power can be made available at the output. A comparable circuit using an SCR instead of a triac can pass only from 0% to 50% of the input waveform, because an SCR permits current to flow in only one direction.

Triacs are pretty efficient devices overall. They are potentially twice as efficient as SCRs (depending on the specific operation of the circuit being used). The reason is that triacs are bidirectional, while SCRs are unidirectional. An SCR must always block half of every ac cycle. This means that half of the input power is wasted. A triac, on the other hand, can pass all of the input cycle, with little inherent waste.

Wasted power in a device such as an SCR or triac is dissipated as heat. While the use of a heat sink is always a good idea for many applications, you can get by with a fairly modest heat sink on a triac. This is true even up to the recommended power limits defined by the manufacturer's specification sheet for the individual unit used. When in doubt, use an external heat sink, just to be on the safe side. Too much heat-sinking will never adversely affect circuit operation.

Although efficient, triacs certainly are not perfect switches. There will be some voltage drop across the device. Typically one or two volts will be dropped across the triac, and this power will be dissipated as heat.

A Triac's operation is very similar to that of an SCR. The gate triggering function works in exactly the same way for both SCRs and Triacs. An SCR is turned off if the current flowing through it reverses polarity (or drops close to zero). But a triac is supposed to conduct current in either direction. You may well be wondering just how we turn the darn thing off once a trigger pulse on the gate has turned it on.

This is a little easier to understand if you remember that the triac is effectively equivalent to a pair of standard SCRs connected back to back. If MT1 is positive with respect to MT2 when the triac is triggered at its gate, SCR A will be turned on. SCR B will stay off, because its anode is negative with respect to its cathode. When the positive signal on MT1 drops below the holding current (I_H) SCR A will be switched off.

If MT2 is positive with respect to MT1, then the entire situation is reversed. Now, when the triac is gated, SCR B will be turned on. SCR A will stay off, because its anode is negative with respect to its cathode. When the positive signal on MT1 drops below the holding current (I_n) SCR B will be switched off.

The triac will be cut off when the current flowing between MT1 and MT2 (in either direction) is very close to zero; that is, if the current is below I_H, with either polarity.

There are four possible combinations for triggering a triac. These are the triggering modes. The differences between the modes lie in the relative polarities of the leads. Normally, all current and voltage polarities for a triac are given with respect to MT1. This convention is followed, simply as a matter of convenience and to avoid unnecessary confusion. There is nothing particularly special about MT1. In fact, MT1 and MT2 are functionally identical. A triac is a nonpolarized device, and therefore it is symmetrical. It works the same way forwards, or backwards. MT1 is just an arbitrarily choosen common reference point.

The triac's four standard triggering modes are:

- MODE A
  MT2 positive with respect to MT1
  Gate pulse positive with respect to MT1

- MODE B
  MT2 positive with respect to MT1
  Gate pulse negative with respect to MT1

- MODE C
  MT1 positive with respect to MT2
  Gate pulse positive with respect to MT1

- MODE D
  MT1 positive with respect to MT2
  Gate pulse negative with respect to MT1

Each of these triggering modes has a different current requirement to trigger the triac. Mode A is the easiest to trigger. It has the lowest current requirement. The current at the gate required to trigger the triac in Mode A is I_{GT} . In Mode B the triac isn't nearly as efficient. A gate current of at least five times I_{GT} is required to trigger the Triac in Mode B. Mode C and Mode D are basically similar to one another. A gate current of about twice I_{GT} is needed to trigger the triac in these Modes, regardless of whether the gate is positive or negative with respect to the MT1 terminal.

DIACS

Triacs often are used with a special two-lead thyristor called the *diac*. A diac can be considered simply as a pair of back-to-back diodes, as illustrated in FIG. 5-7. Current of either polarity can flow through one of the two diodes, so the pair, as a unit, is bidirectional. Each diode is individually unidirectional, as all diodes are. If an ac signal is applied across the pair, diode A will conduct during the positive half-cycles, while diode B will take over the conduction on the negative half-cycles. Each diode passes half of the complete cycle. Together, they pass the

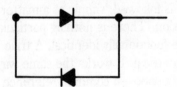

Fig. 5-7. Functionally, a diac resembles a pair of back to back diodes.

entire waveform cycle. Because of this bidirectional conduction, you could call a diac a two-way diode. The back-to-back diode model is suggested strongly by the schematic symbols generally used to represent diacs, which are shown in FIG. 5-8.

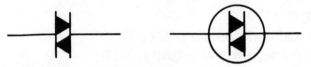

Fig. 5-8. The schematic symbols used to represent diacs.

Diacs are usually, though not always, used in conjunction with triacs. This is because both diacs and triacs are bidirectional devices. An SCR, being a unidirectional device, cannot take much advantage of the diac's special capabilities. A diac may be used with an SCR, but in most applications, there really isn't much reason to do so.

In SCR ac power controller circuits, the SCR often is triggered via a neon lamp. This triggering method also may be used with a triac, but the neon lamp can trigger the thyristor (SCR or triac) only during the rising portion of the positive half-cycle of the input waveform. In this case, the triac offers no particular advantage over the SCR. Both components will work in exactly the same way in circuits of this type.

If the neon lamp is replaced with a Diac, the triac will be *bilaterally triggered*; that is, it will be triggered on both positive half-cycles and negative half-cycles. This allows the circuit to operate more efficiently.

You should remember from the preceding section that while a triac can be triggered with either a positive or negative gate pulse, a larger negative pulse is required to turn on the device. This means that the triac will be turned on later in each negative half-cycle than it is in each positive half-cycle.

Using a diac neatly compensates for the differences in the positive and negative switching voltages required by the triac. The diac doesn't care at all if it is being triggered positively or negatively. In either case, it will put out a sharp pulse that is fed to the gate of the triac. The triac will switch on at approximately the same point in the negative half-cycles as it does in the positive half-cycles.

The diac is nothing more than a gateless triac. The two remaining terminals on a diac are labelled in the same way as on a triac, MT1 and MT2. Sometimes the terminals are labelled "anode 1," and "anode 2," but this is a little misleading. There is no difference between these two terminals. They are entirely interchangable.

The diac may be wired into a circuit in either direction. It is not a polarized device. This makes it a rather unusual semiconductor component. Most semiconductor devices have very definite polarity requirements, and can be accidentally wired into a circuit backwards. But it is impossible to hook up a diac backwards. It will work fine, no matter which way it is facing in the circuit.

QUADRACS

From time to time, most electronics hobbyists will encounter SCRs, triacs and diacs. Actually, SCRs are becoming fairly common. Even triacs and diacs are being employed more in many modern circuits. There is a close relative to these devices which most hobbyists are completely unfamiliar with. This is the *quadrac*.

Like most other thyristor devices, the diac is used primarily for switching applications. Being a switching device that can conduct current in either direction, the diac often is called a *bilateral switch*. The diac is a nonpolarized component.

While rarely used in hobbyist projects, quadracs are found occasionally in some commercial electronic equipment. It is useful to have some familiarity with this device, even though you never may use it yourself. Therefore, this brief chapter will offer you a quick introduction to the quadrac.

Since a diac has two leads and a triac has three leads, it might seem logical that a quadrac should have four leads. Logical or not, this is not the case. A quadrac has only three leads, just like a triac. I don't know the origin of the name, but I suspect it might be from the fact that a quadrac goes "one step further" than a triac, or that it contains four devices.

The schematic symbol for a quadrac is shown in FIG. 5-9. Having gotten this far in this chapter, you really already know just about everything there is to know about the quadrac. As the schematic symbol suggests, a quadrac is a triac and a diac in a single housing. This is a very logical arrangement, because these two component types are used frequently together. The internal diac is hard-wired to the gate of the internal triac. Sometimes a quadrac is referred to as a *triac with trigger*.

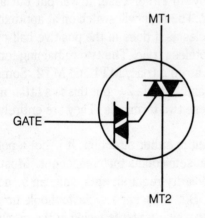

Fig. 5-9. A quadrac contains a triac and a diac in a single housing.

The three leads to a quadrac are usually labeled the same as the leads for a triac (Gate, MT1, and MT2). In some sources, the main terminals (MT1 and MT2) may be called anode 1 and anode 2, or high and common. Whatever names are used, these terminals are equivalent to the MT1 and MT2 terminals of a triac. A quadrac can be used in place of any triac and diac combination, with no other changes in the external circuitry.

You may be wondering why the quadrac isn't used more widely by electronics hobbyists. It would seem handy in reducing the parts count and complexity of many circuits. Aside from the general hobbyist's lack of familiarity with this component, the main problem is availability and cost. These two aspects of the problem are tightly interwoven. Quadracs aren't generally available to the electronics hobbyist. They are carried by very few electronics parts stores or mail-order houses. This is somewhat a chicken and egg syndrome. The quadrac is not stocked widely because it is not popular and doesn't sell well. It is not popular largely because of its limited availability and unfamiliarity.

It might seem that a Quadrac would offer a considerable cost advantage. After all, one component replacing two should cut the cost of a circuit, shouldn't it? It probably should, but in this case, it simply doesn't. Quadracs are sold for reasonable prices in large quantities, so they are appealing to manufacturers, especially in products where size is an important consideration. Unfortunately, due to the limited popularity of quadracs among hobbyists, when you can find them at all, they tend to be quite expensive in single quantities. In most cases, it will be cheaper and easier to use a separate triac and diac instead of a quadrac.

There isn't likely to be a sudden large demand for this type of device because a quadrac doesn't offer that much of an advantage over separate triac/diac combinations. The price/availability problems of the quadrac are not likely to change in the near future.

6
Multivibrators

THERE IS A SPECIAL CLASS OF SWITCHING CIRCUITS KNOWN AS *MULTIVIBRA-tors*. Multivibrator circuits frequently are called *pulse circuits*.

A multivibrator has only two possible output states. Either the output state is high, or it is low. There are no other possibilities, or intermediate values. The output signal can be switched from one state to the other (high to low, or low to high) with (theoretically) no transition time between the two states.

There are three different types of multivibrator circuits. They are identified as follows:

- Monostable multivibrators
- Bistable multivibrators
- Astable multivibrators

Each of these basic multivibrator circuits will be discussed in this chapter.

MONOSTABLE MULTIVIBRATORS

A monostable multivibrator, as its name suggests, has one stable output state. The opposite output state can be generated, but it is unstable. The output at some point will revert automatically to its stable state. The stable state may be either high or low, depending on the specific circuit design. All monostable multivibrator circuits have an input connection which is used to trigger the multivibrator into its opposite, unstable state. In the following discussion, assume that the stable state is low.

When power first is applied to the monostable multivibrator circuit, its output will be low. The output will remain low indefinitely, unless the circuit is acted upon by an external signal. When a suitable pulse is received at the circuit's trigger input, the output will switch over to the high state. This high output will be held for a fixed period of time. The time period is determined by certain component values within the monostable multivibrator circuit. After the preset time period has passed, the multivibrator's output will jump back to its normal low state. The output will be held low until another trigger pulse is detected by the circuit, or power is disconnected.

Typical input and output signals for a monostable multivibrator are illustrated in FIG. 6-1. The length of the output pulse is determined by component values in the circuit. This means that the length of the output pulse will be constant, regardless of the length of the input pulse. For this reason, monostable multivibrator circuits often are called *timers*.

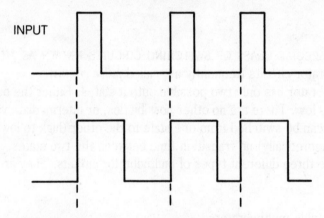

INPUT

Fig. 6-1. Typical input and output signals for a monostable multivibrator.

In the vast majority of practical applications, the output pulse will be significantly longer than the input pulse. Another common name for a monostable multivibrator circuit is a *pulse stretcher*. The input pulse is made effectively longer (it is stretched) at the output.

The 555 Timer IC

There are many possible approaches to designing a monostable multivibrator circuit. One of the easiest and most efficient for many applications is to use a 555 timer IC. The 555 is an integrated circuit especially designed for monostable and astable (discussed later in this chapter) multivibrator applications.

As the pin-out diagram of FIG. 6-2 demonstrates, the 555 is not a particularly complex device. In its most common form, the 555 is supplied in a standard 8 pin DIP housing. Some of the pin labels may seem rather more exotic than they really are. For that reason, the 555 will be examined pin by pin.

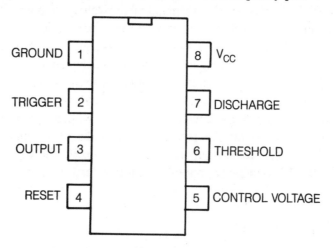

Fig. 6-2. The 555 timer IC is used in many multivibrator circuits.

Pin #1 is the ground terminal for the device. The 555 must be used with a negative ground. No below ground (more negative) voltages should ever be applied to this chip. The voltage on pin #1 should be the lowest (most negative) voltage applied to any of the integrated circuit's pins.

Pin #2 is the trigger input. This is the control input for monostable multivibrator applications. Normally this terminal is held at a constant value that is at least one third of the source voltage. If the trigger voltage drops below the one third source point, it will trigger the 555, and the timing cycle will be initiated. Notice that the 555 timer requires a negative going trigger pulse. The circuit is triggered when the input drops from high to low.

Pin #3 is the usual output of the 555 timer circuit. Some specialized applications may call for alternative output points.

Pin #4 is labelled *reset*. As the name suggests, the signal on this terminal is used to return the device to its original rest (stable) state at the end of the timing cycle.

Pin #5 is left unused in most circuit applications, but in some special cases, this pin can be extremely useful. It is a control voltage input. The trigger voltage point can be determined by an external voltage applied to this terminal, in place of the normal one third source voltage trigger point. When the voltage control mode is not used, pin #5 should be connected to ground through a 0.01 μF capacitor. This is not always necessary, but it is a cheap insurance against possible stability problems in some circuits. When in doubt, go ahead and use the capacitor. It never will do any harm.

Pin #6 is the threshold input. The voltage on this terminal tells the timer when to end its timing cycle. In virtually all 555 circuits, a timing resistor is connected from pin #6 to the positive terminal of the voltage supply. The value of this resistor is one of the factors determining the timing period of the circuit.

Pin #7 is the discharge pin. This terminal also is used to determine the length of the timing cycle. A timing capacitor is normally connected between pin #7 and ground. The value of this capacitor combines with the value of the timing resistor (connected to pin #6) to determine the circuit's timing period.

Finally, pin #8 is the chip's power supply pin. The positive terminal of the voltage source is connected to this pin. In schematic diagrams this pin usually is labeled V+ or V_{CC}. The voltage supply for a 555 timer IC should be between 5 and 15 volts, with 15 volts being generally preferred for the best results in most applications.

The 556 Dual Timer IC

More complex circuits often use the 556 dual timer IC, which is shown in FIG. 6-3. The 556 contains two independent 555-type timer circuits in a single

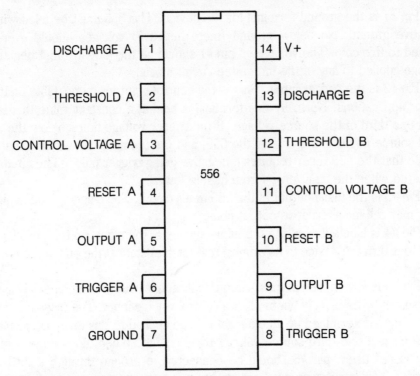

Fig. 6-3. The 556 IC contains two independent 555 type timers in a single package.

14-pin DIP housing. All functions are brought out separately for each of the internal timers. The only pins shared are the power supply leads (V+ and ground).

A 556 dual timer IC can be substituted for two separate 555 timer ICs, or vice versa, with no differences in circuit operation. When making such substitutions, just be careful to match up the correct pin numbers. Refer back to FIG. 6-2 and FIG. 6-3.

The 555 Monostable Multivibrator Circuit

The basic 555 monostable multivibrator circuit is illustrated in FIG. 6-4. As you can see, this circuit is quite straightforward. A negative (high and low) input pulse is used to trigger the circuit.

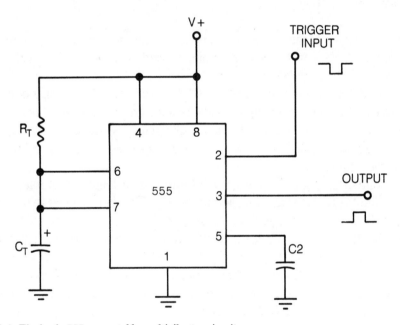

Fig. 6-4. The basic 555 monostable multivibrator circuit.

The normal (stable) output state of this circuit is low. When the timer is triggered, the output goes high for a specific period of time determined by the values of the timing resistor (R_T) and the timing capacitor (C_T), according to this simple formula:

$$T = 1.1 \ (R_T C_T)$$

where T is the time period of the output pulse in seconds, R_T is the resistance

73

between pin #6 and V_{CC} in megohms (1 Megohm = 1,000,000 ohms), and C_T is the capacitance between pin #7 and ground in microfarads (μF).

For the most reliable operation, the values of the timing components (Rt and Ct) should be kept within certain limits. The suitable range of values for the timing resistor (Rt) is as follows:

Minimum	0.01 Megohm	10,000 ohms	10KΩ
Maximum	10 Megohms	10,000,000 ohms	10MΩ

Similarly, the timing capacitor (Ct) should be kept within the following range of values:

Minimum	0.0001 μF	100pF
Maximum	1000 μF	

Keeping the values of the timing components within these ranges is not much of a restriction. A very wide range of timing periods can be created with timing components in these ranges.

Using the timing equation, $T = 1.1(R_T C_T)$, the acceptable range of component values can produce timing periods ranging from 0.0000011 second (1.1 microseconds) to 11,000 seconds (about 183 minutes, or just over 3 hours). Clearly, the 555 is a very wide range device, especially considering how inexpensive the chip is.

The output timing pulse always will be the same length, regardless of the length of the triggering signal (applied to pin #2). This is providing that the trigger signal lasts a shorter time than the desired output pulse. If the trigger pulse is too long, the 555 may be retriggered automatically as soon as it times out (completes its timing period).

To get a good feel for the basic 555 monostable multivibrator circuit, here is a quick design example. Assume that you need a monostable multivibrator circuit with an output pulse that lasts 17 seconds. First, select a likely value for the timing capacitor (try a 22 μF capacitor). Next, rearrange the timing equation to solve for the unknown timing resistor value:

$$R_T = T/1.1C_T$$

Plugging in the time and capacitor values for the example, and solving, finds a timing resistor with a value of:

$$R_T = 17/(1.1 \times 22)$$
$$= 17/24.2$$

$$= 0.70 \text{ megohms}$$
$$= 700,000 \text{ ohms}$$

The closest standard resistance value would be a 680k resistor. Such rounding off usually will be permissible in most practical applications. Component tolerances usually result in at least as much error anyway.

In precision applications, use a trimpot for the timing resistor, and carefully adjust the trimpot for the exact timing period desired. This technique also will compensate for any tolerance variations in the value of the timing capacitor.

In such design equations, it is usual to start with an arbitrary capacitor value and then solve for the resistance because resistors come in a wider series of values and are easier to adjust for unusual values than are capacitors.

Transistor Monostable Multivibrator Circuit

A monostable multivibrator circuit built around discrete transistors is shown in FIG. 6-5. Ordinarily, when there is no signal at the circuit's input, transistor Q1 is held on because of the biasing current to its base through resistor R2 from the supply voltage ($+V_{CC}$). At the same time, a negative voltage ($-V_{BB}$) is applied to the base of transistor Q2 through resistor R6. The negative voltage ($-V_{BB}$) must be less than the circuit's positive supply voltage ($+V_{CC}$). This negative bias voltage keeps the second transistor in the cutoff state.

Because transistor Q2 is off, the circuit's output is high. Resistor R5 pulls the output up to a value just under the circuit's positive supply voltage ($+V_{CC}$). These conditions remain unchanged while power is applied to the circuit and there is no signal at the input.

If a positive pulse is applied to the circuit input, nothing will happen, the circuit will continue to hold its normal, untriggered conditions. The output stays low. Positive input signals are ignored by the circuit because they do not change the biasing on transistor Q1.

A sufficient negative pulse at the input, however, will overcome the positive bias on the base of transistor Q1, cutting this device off. The collector voltage of Q1 will jump up to a value near the supply voltage ($+V_{CC}$). This positive voltage is now fed to the base of transistor Q2 through resistors R3 and R4. The $+V_{CC}$ voltage has a higher potential than the $-V_{BB}$ voltage, so the negative bias on transistor Q2 is overcome, and this unit is now switched on. The circuit's output goes low.

Even if the input pulse is now removed, transistor Q1 will remain off until capacitor C1 has sufficient time to discharge through resistor R2. Once the capacitor has discharged below a critical level (one-third of its maximum voltage), transistor Q1 is turned back on, and transistor Q2 reverts to the cutoff

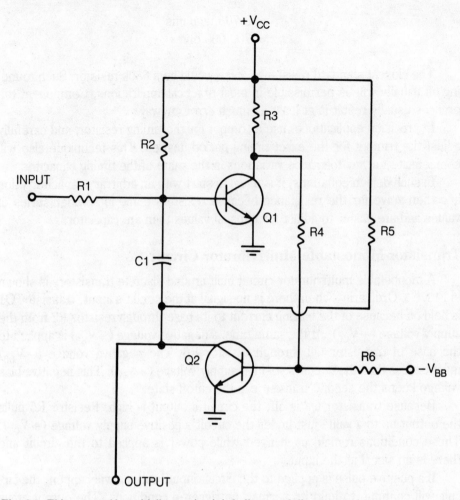

Fig. 6-5. This monostable multivibrator circuit is built around discrete transistors.

state, returning to its original circuit conditions. The circuit automatically reverts to its normal, stable state until another negative input pulse is received at the circuit input.

Resistor R2 and capacitor C1 control the timing period of this monostable multivibrator circuit. The time required for the capacitor to discharge enough to turn on transistor Q1 can be found with this fairly simple formula:

$$T = 0.69R_2C_1$$

where R_2 is the resistance in ohms and C_1 is the capacitance in farads.

A few examples will help make things clearer. Assume the following component values:

R2   220K   220,000 ohms
C1   0.05 μF   0.0000005 farad

Using these component values, the circuit's timing period works out to:

$$T = 0.69 \times 220000\Omega \times 0.00000005f$$
$$= 0.00759 \text{ second}$$
$$\cong 8 \text{ mS}$$

If you leave the value of resistor R2 alone, but increase the value of capacitor C1 to 25 μF (0.000025 farad), the circuit's timing period becomes:

$$T = 0.69 \times 220000\Omega \times 0.000025f$$
$$= 3.795 \text{ seconds}$$
$$\cong 4 \text{ seconds}$$

For a third and final example, retain a value of 25 μF for capacitor C1, but increase the resistance of resistor R2 to 680K (680,000 ohms). With these changed component values, the circuit's timing period works out to:

$$T = 0.69 \times 680,000\Omega \times 0.000025f$$
$$= 11.73 \text{ seconds}$$
$$\cong 12 \text{ seconds}$$

Increasing either the resistance or the capacitance (or both) will increase the length of the timing period. Similarly, decreasing either (or both) of these component values will shorten the circuit's timing period.

As with all monostable multivibrator circuits, the length of the short input pulse has no effect on the length of the output pulse. Once the timing cycle has been started, its period will be determined solely by the RC time constant of the circuit itself. External signals are irrelevant.

BISTABLE MULTIVIBRATORS

The prefix *mono-* means one, so a monostable multivibrator has one stable output state. The prefix *bi-* means two, so a *bistable multivibrator* has two stable output states.

In a bistable multivibrator circuit, both of the two available output states (high and low) are stable, and either may be held indefinitely. Each time the circuit is triggered (receives an external pulse), the output state reverses itself. If the output is initially low, the first input pulse will cause the output to jump to high. A second input pulse will bring the output state back down to low. Either output state can be held as long as power is applied to the circuit, and no input pulse is detected.

A bistable multivibrator circuit is essentially a very simple (one bit) memory circuit. In effect, the circuit remembers the output state from the last time it was triggered. Bistable multivibrators sometimes are called *flip-flops*, because of the back and forth nature of their input and output signals. Typical input and output signals for a bistable multivibrator circuit are illustrated in FIG. 6-6.

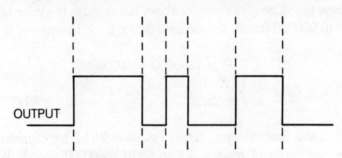

Fig. 6-6. Typical input and output signals for a bistable multivibrator.

If the input signal is a square wave, the output signal also will be a square wave with a frequency which is exactly one half of that of the input signal. For this reason, bistable multivibrator circuits also are referred to as *frequency dividers*.

Most bistable multivibrator circuits have two inputs. One of these inputs, which often is labeled *set*, forces the circuit's output to a high level. The other input, which generally is called *clear* or *reset*, drives the circuit's output to a low level. This type of bistable multivibrator circuit is known as the *set-reset* flip-flop, or the *RS* flip-flop.

Another type of bistable multivibrator circuit, the *D-type* flip-flop, controls both high and low output states with a single input signal. The circuitry for the D-type flip-flop is somewhat more complicated than for an RS flip-flop, so details on that type of circuit will not be discussed.

Two outputs often are included in bistable multivibrator circuits. The main output usually is labeled Q, and the other output is identified as $\overline{Q}$(not Q). These two outputs are complementary; that is, they are always in opposite states. If Q is high, then $\overline{Q}$ is low, and if Q is low, then $\overline{Q}$ is high. There are no other possible output combinations.

Transistor Bistable Multivibrator Circuit

A fairly typical bistable multivibrator circuit built around a pair of transistors is illustrated in FIG. 6-7. This particular circuit is an RS flip-flop. The two halves of the circuit are mirror images of each other. Similar components must have equivalent values. That is:

$$Q1 = Q2$$
$$R1 = R4$$
$$R2 = R5$$
$$R3 = R6$$

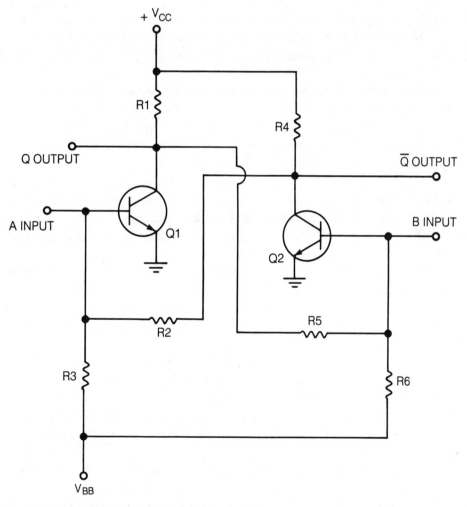

Fig. 6-7. A fairly typical bistable multivibrator circuit.

In this circuit, one of the transistors always will be in cutoff (off), and the other transistor always will be in saturation (on). The input pulses reverse the states of the two transistors, and their associated outputs (Q and $\overline{Q}$).

When power first is applied to this circuit, one of the transistors inevitably will start to conduct a little faster than its partner. That transistor will go into saturation, and the slower transistor will be cutoff. It really doesn't matter which is which, and it likely will be different each time the circuit is used.

For purposes of discussion here, assume that initially transistor Q1 is saturated (on), and transistor Q2 is in cutoff (off). At this point, the collector of Q1 (Q output) is near ground potential (low), and the collector of transistor Q2 (the complementary $\overline{Q}$ output) is at a voltage close to V_{CC} (high).

Assuming that there is no input signal (at either input), the base of Q2 is fed Q1's collector signal (ground potential) through resistor R5, and bias voltage $-V_{BB}$ through resistor R6. Since the voltage of the collector of transistor Q1 is currently zero, the overall voltage applied to the base of transistor Q2 is negative, holding this device in cutoff.

Similarly, still assuming there is no input signal, the base of Q1 is fed Q2's collector signal through resistor R2, and bias voltage $-V_{BB}$ through resistor R3. Since the potential of Q2's collector is currently close to $+V_{CC}$, which is significantly larger than $-V_{BB}$, the overall voltage on the base of transistor Q1 is a relatively large positive value. This keeps Q1 in saturation.

As you can see, this circuit is in a stable (unchanging) condition, as long as no input signal is received (and power to the circuit is not interrupted). If a positive voltage pulse is now applied to input A, nothing will happen, since this pulse will be added to the already positive voltage on the base of transistor Q1. The current output states continue to be held.

On the other hand, if a sufficiently positive pulse is applied to input B, it will overcome the negative bias on the base of transistor Q2. This will drive Q2 into saturation. Its collector voltage almost instantly drops to zero (ground potential), so the controlling voltage on the base of Q1 goes negative, cutting this device off.

Now, there is the opposite condition. The collector of transistor Q1 is almost equal to $+V_{CC}$ (output Q is high), and the collector of Q2 is close to ground potential (complementary output $\overline{Q}$ is low). This new state of affairs will remain constant, even if the pulse voltage at input B is now removed. If another positive pulse is applied to input B, nothing will happen, since Q2 is already in saturation.

If a positive pulse now is applied to input A, transistor Q1 will be turned on, and transistor Q2 will be turned off, returning to the original starting condition. As long as power is applied continuously, the circuit will remember which input was last activated.

The inputs and outputs for this bistable multivibrator circuit can be summarized as follows:

Inputs		Outputs	
A	B	Q	$\overline{Q}$
L	L	No Change	
H	L	L	H
L	H	H	L
H	H	Disallowed State	

L represents a low signal, and H indicates a high signal.

One possible input combination (a positive pulse simultaneously at both inputs) results in a disallowed state. The output states will be impossible to predict, and in some cases the circuitry may be damaged. Such disallowed states must be avoided in the operation of the circuit. Timer ICs, such as the 555 and the 556, discussed earlier in this chapter, generally are not suitable for use in bistable multivibrator circuits.

ASTABLE MULTIVIBRATORS

The third and final type of multivibrator circuit is the *astable multivibrator*. Remember that a monostable multivibrator has one stable output state, and a bistable multivibrator has two. The prefix *a-* means no, or none. In an astable multivibrator circuit, neither output state is stable. The output keeps switching back and forth between the low and high states. No input signal is required.

A typical output signal from an astable multivibrator circuit is illustrated in FIG. 6-8. Essentially, this is just a square wave of a specific frequency (determined by component values within the circuit). An astable multivibrator circuit is a form of oscillator.

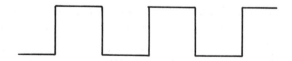

Fig. 6-8. The output of a symmetrical astable multivibrator is simply a square wave.

Some astable multivibrator circuits are designed to be unbalanced. That is, the output will hold one output state longer than the other. This gives us an output that is an asymmetrical rectangle wave, like one of the signals shown in FIG.

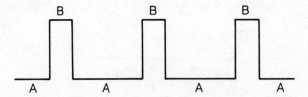

Fig. 6-9. The output of an asymmetrical astable multivibrator is simply a rectangle wave.

6-9. Unequal timing component values are used in the two halves of the circuit to create this effect. Otherwise, the same operating principles apply.

An astable multivibrator circuit has two timing periods, usually (though not always) set by separate RC (resistor—capacitor) combinations. For convenience, just call the two timing periods A and B.

Assume that initially the output is in the low state. Timing period A is in effect. After this part of the circuit times out, the output goes high, and timing period B begins. After timing period B ends, the output switches back to the low state, and timing period A starts again. This back and forth process continues as long as power is applied to the circuit. No external input signal is required. The total cycle time is the sum of the low time (A) and the high time (B):

$$T = A + B$$

For many applications, it is more convenient to describe the output switching rate in terms of frequency, instead of time. This is easy enough to do, since frequency is simply equal to the reciprocal of the total cycle time:

$$F = 1/T$$

In working with rectangle waves, you often will encounter the term *duty cycle*. This is a measurement of the signal's symmetry. The duty cycle is the ratio of the high time (B) to the total cycle time. For a true square wave, the high time is exactly equal to the low time. The output signal is high for exactly one-half of each complete cycle. The duty cycle in this case is 1:2.

If the output signal is high for one-third of each complete cycle, the duty cycle would be 1:3. In practical circuits, the times may not be evenly divisible. Fractional duty cycles, such as 1:4.37 or 1:3.88, are not uncommon. Such duty cycle values are often a little awkward to work with. In most applications, it is permissible to round off the duty cycle values to the nearest whole numbers. For these examples, the rounded-off duty cycles would both be 1:4.

All ac waveforms, except the sine wave, are made up of multiple-frequency components. The cycle repetition rate is the *fundamental frequency* of the signal.

This fundamental frequency will be the perceived pitch of the tone in most cases. Generally speaking, the fundamental frequency will be the strongest (highest amplitude) frequency component in the signal.

In addition to the fundamental frequency, there are secondary frequency components known as *harmonics*. A harmonic is a frequency component with a frequency that is an exact whole number multiple of the fundamental frequency. For example:

Fundamental Frequency	100 Hz
Second Harmonic	200 Hz
Third Harmonic	300 Hz
Fourth Harmonic	400 Hz
Fifth Harmonic	500 Hz
Sixth Harmonic	600 Hz
Tenth Harmonic	1000 Hz
etc.	

In most waveforms, the higher harmonics are weaker (lower in amplitude) than the lower harmonics. The fundamental frequency usually has the strongest amplitude of all of the frequency components in the waveform.

Many practical waveforms are *incomplete*, that is, they do not include all of the available harmonic components. This is true of rectanglular waves. The duty cycle of the waveshape will determine which harmonics are omitted from the signal. All harmonics which are exact, whole number multiples of the second (larger) number in the duty cycle ratio are left out. For example, a symmetrical square wave has a duty cycle of 1:2. All harmonics that are multiples of two are omitted from the signal's makeup. You could, for instance, break down a 250 Hz square wave like this:

Fundamental	250 Hz
Third Harmonic	750 Hz
Fifth Harmonic	1250 Hz
Seventh Harmonic	1750 Hz
Ninth Harmonic	2250 Hz
etc.	

If the duty cycle of the rectangle wave was 1:3, then every third harmonic would be left out, and the following frequency components would be included in the total signal:

Fundamental	250 Hz
Second Harmonic	500 Hz

Fourth Harmonic	1000 Hz
Fifth Harmonic	1250 Hz
Seventh Harmonic	1750 Hz
Eighth Harmonic	2000 Hz
Etc.	

Transistor Astable Multivibrator Circuit

A fairly typical transistor based astable multivibrator circuit is illustrated in FIG. 6-10. Notice the similarities among this circuit, the monostable multivibrator circuit, and the bistable multivibrator circuits presented earlier in this chapter. This circuit, like the bistable multivibrator circuit discussed earlier, can be divided into two sections, which are mirror images of each other. One section includes transistor Q1, resistors R1 and R2, and capacitor C1. The second section is comprised of transistor Q2, resistors R3 and R4, and capacitor C2.

When power first is applied to this circuit, one of the two transistors will start to conduct a little faster than the other transistor. It doesn't matter which

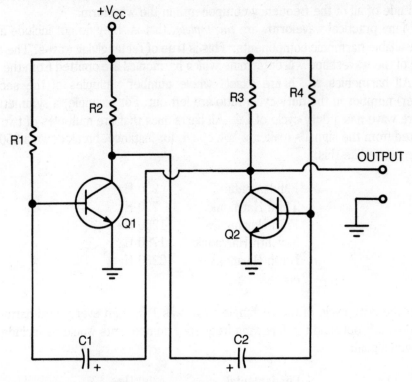

Fig. 6-10. A simple transistorized astable multivibrator circuit.

is which. The faster transistor will go into saturation, and the slower transistor will go into cutoff.

Assuming that initially transistor Q1 is on, and transistor Q2 is off. Under these conditions, the collector voltage of Q1 will be near ground potential (0 volts), and the collector voltage of Q2 will be just a little under the circuit's supply voltage ($+V_{CC}$). Capacitor C1 is charged up to a value just under V_{CC} by the collector voltage of Q2. The polarity of this charge will be as indicated in the schematic diagram (FIG. 6-10.).

During the previous half-cycle, capacitor C2 had been charged with the polarity shown in the diagram. As capacitor C1 is now charged up, capacitor C2 starts to discharge. When C2 has been discharged, a base current flows through transistor Q2 due to the current flowing through resistor R4 from V_{CC}. This positive base current turns Q2 on, pulling its collector and the positive end of capacitor C1 down to ground potential.

Because of the polarity of the charge on C1, the base of transistor Q1 is negative, cutting this transistor off. During the time Q1 is off, its collector is at V_{CC} charging capacitor C2 with the polarity shown. As capacitor C2 is being charged up, capacitor C1 is being discharged through resistor R1, and the whole process is reversed.

For a square wave output (duty cycle of 1:2), the two halves of the circuit should have symmetrical component values. That is:

$$Q1 = Q2$$
$$C1 = C2$$
$$R1 = R4$$
$$R2 = R3$$

The length of the entire cycle would be equal to:

$$T = 1.38R_1C_1$$

Since frequency is the reciprocal of the total cycle time, the formula can be rearranged as follows:

$$F = 1/T$$
$$= 1/(1.38R_1C_1)$$

As a typical design example, assume you are using the following component values:

$$R1 = R4 = 27K\Omega \qquad 27,000 \text{ ohms}$$
$$C1 = C2 = 0.33\mu F \qquad 0.00000033 \text{ farad}$$

Using these component values, the output frequency works out to a value of:

$$F = 1/(1.38 \times 27000 \times 0.00000033)$$
$$= 1/0.0122958$$
$$= 81 \text{ Hz}$$

If unequal component values are used in the two halves of the circuit, the output waveform will not be symmetrical. Some duty cycle other than 1:2 can be set up in this manner. The equations become only slightly more complicated in this case. The high level output time (T_H) equals:

$$T_H = 0.69 R_4 C_2$$

The low level output time (T_1) is:

$$T_1 = 0.69 R_1 C_1$$

The total cycle time (T_T) is the sum of the low time (T_1), and the high time (T_H):

$$T_T = T_H + T_1$$
$$= 0.69 R_4 C_2 + 0.69 R_1 C_1$$

The output frequency is still the reciprocal of the total cycle time:

$$F = 1/T_T$$
$$= 1/(T_H + T_1)$$
$$= 1/(0.69 R_4 C_2 + 0.69 R_1 C_1)$$

As a typical design example, use the following component values in the astable multivibrator circuit:

R1	= 1KΩ	1000 ohms
R4	= 4.7KΩ	4700 ohms
C1	= 0.1 μF	0.0000001 farad
C2	= 0.033 μF	0.000000033 farad

Using these component values, the high level output time works out to:

$$T_H = 0.69 \times 4700 \times 0.000000033$$
$$= 0.000107 \text{ second}$$
$$\cong 0.11 \text{ millisecond}$$

Similarly, the low time is:

$$T_1 = 0.69 \times 1000 \times 0.0000001$$
$$= 0.000069 \text{ second}$$
$$= 0.069 \text{ millisecond}$$

The total cycle time in this circuit is therefore equal to:

$$T_T = T_H + T_1$$
$$= 0.11 + 0.069$$
$$= 0.179 \text{ millisecond}$$

You can now figure out the duty cycle, simply by comparing the high output time (T_H) to the total cycle time (T_T):

$$T_H : T_T$$
$$0.11 : 0.179$$
$$1 : 1.63$$

Finally, the output frequency for this particular circuit is:

$$F = 1/T_T$$
$$= 1/0.000179$$
$$= 5587 \text{ Hz}$$

This astable multivibrator circuit can cover a wide range of output frequencies by the proper selection of component values.

A secondary output, with a signal that is 180° out of phase with the main output, can be pulled easily from this circuit. This secondary output can be tapped off from the collector of transistor Q1. When one output goes high, the other will go low, and vice versa. The two outputs always will be in opposite states. There are many practical applications in which such complementary outputs can come in quite handy.

555 Astable Multivibrator Circuit

Timer ICs, such as the 555 and the 556 (discussed earlier in this chapter), can be used to create astable multivibrator circuits. These circuits are quite simple, with a relatively low component count, and fairly easy design equations. The basic 555 astable multivibrator circuit is illustrated in FIG. 6-11. Compare this to

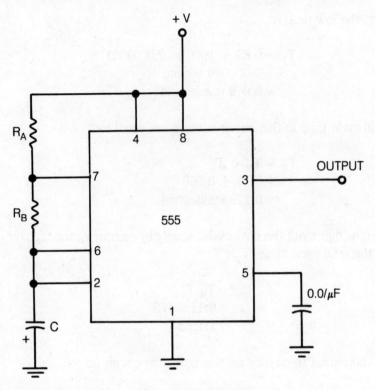

Fig. 6-11. The 555 timer used in an astable multivibrator circuit.

the 555 monostable multivibrator circuit presented earlier in this chapter (FIG. 6-4).

The main difference between this astable multivibrator circuit and the earlier monostable multivibrator circuit is that the timing resistor (R_T) has been split into two separate resistors (R_A and R_B), and there is no input for an external trigger signal. An astable multivibrator circuit is self-triggering.

The time the output from this circuit is at its high voltage state is determined by capacitor Ct and resistors Ra and Rb, according to this formula:

$$T_H = 0.693 \times (R_A + R_B) \times C_T$$

The time the output is in its low (or grounded) state depends on only capacitor C_T and resistor R_B. Resistor R_A is now ignored. The formula for determining the low time is:

$$T_1 = 0.693R_BC_T$$

The total time of the complete cycle (T_T) is the sum of the high time (T_H) and the low time (T_1):

$$T_T = T_H + T_1$$

These three equations can be combined and rewritten in this form:

$$T_T = 0.693 \times (R_A + 2R_B) \times C_T$$

The duty cycle of the output waveform will depend on the relative values of resistors R_A and R_B. It is not possible to achieve a 1:2 duty cycle (true square wave) with this circuit. The frequency of the output signal is the reciprocal of the total cycle time:

$$F = 1/T_T$$

Combining the various timing equations, you can rewrite the frequency equation as:

$$F = 1.44/((R_A + 2R_B)C_T)$$

As an example, assume you are working with an astable multivibrator circuit with the following component values:

C_T = 0.1 μF   0.0000001 farad
R_A = 22KΩ 22,000 ohms
R_B = 10KΩ 10,000 ohms

In this sample circuit, the high time works out to:

$$
\begin{aligned}
T_H &= 0.693 \times (22000 + 10000) \times 0.0000001 \\
&= 0.693 \times 32000 \times 0.0000001 \\
&= 0.0022 \text{ second} \\
&= 2.2 \text{ mS}
\end{aligned}
$$

The low time is equal to:

$$
\begin{aligned}
T_L &= 0.693 \times 10000 \times 0.0000001 \\
&= 0.000693 \text{ second} \\
&\cong 0.7 \text{ mS}
\end{aligned}
$$

This gives us a total cycle time of:

$$T_T = 0.0022 + 0.0007$$
$$= 0.0029 \text{ second}$$

Taking the reciprocal, we can find the output frequency for this sample circuit:

$$F = 1/T_T$$
$$= 1/0.0029$$
$$= 345 \text{ Hz}$$

Or, using the alternate frequency equation, we get a final value of:

$$F = 1.44/((22000 + (2 \times 10000)) \times 0.0000001)$$
$$= 1.44/((22000 + 20000) \times 0.0000001)$$
$$= 1.44/(42000 \times 0.0000001)$$
$$= 1.44/0.0042$$
$$= 343 \text{ Hz}$$

The small error between the two results is due to rounding of values in the equations. The difference is negligible. This basic 555 astable multivibrator circuit is capable of a very wide range of possible output frequencies.

7
Digitally Controlled Bilateral Switches

DIGITAL SIGNALS ARE BEING USED IN MORE AND MORE APPLICATIONS THESE days. In a sense, all digital circuits are switching circuits, but an in depth study of digital circuitry is beyond the scope of this book. Only analog switching functions will be emphasized here. In this chapter, some specialized devices designed to switch analog (or digital) signals under the control of a simple digital signal will be looked at.

WHAT IS A BILATERAL SWITCH?

A digitally controlled switching unit is known as a *bilateral switch*. A conceptual diagram of a bilateral switch is shown as FIG. 7-1. Notice that there are three leads. Two of the leads are the opposite ends of a SPST switch. Like any standard SPST switch, the signal may pass through the switch in either direction, which is why the term *bilateral* is used for this device. The switch is nonpolarized, and the two switch terminals are fully interchangeable. It doesn't matter which is the input and which is the output. The signal applied across the switch can be any analog (ac or dc) or digital signal that does not exceed the maximum voltage or current ratings of the device.

The third terminal on the bilateral switch is the interesting one. This is the control terminal. A logic 0 (low) is applied to this terminal to open up the switch, while a logic 1 (high) closes it. This terminal is strictly an input. The signal presented to the control terminal must be an appropriate digital signal.

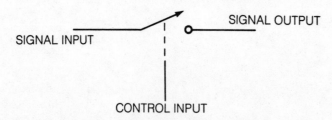

Fig. 7-1. A bilateral switch operates under digital control.

The bilateral switch can be used in just about any switching application, just as if it was a regular mechanical switch. The only difference is that the switching function is under digital control. Practical bilateral switches are usually CMOS ICs.

THE CD4066 QUAD BILATERAL SWITCH

By far, the most common type of bilateral switch is the CD4066. This unit is a CMOS IC containing four independent bilateral switches in a single 14 pin DIP housing. The pin-out diagram for the CD4066 is shown in FIG. 7-2. Remember that the inputs and outputs are completely interchangeable.

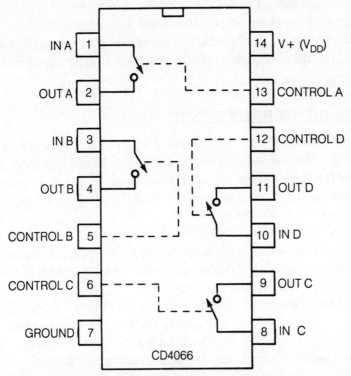

Fig. 7-2. The CD4066 contains four SPST bilateral switches in a single package.

The four SPST switches in the CD4066 can be used separately, or they may be ganged together in any desired configuration. For example, if the IC is wired as shown in FIG. 7-3, it will simulate the action of a DPST switch. Two of the bilateral switches are operated from the same digital control signal, so they always operate in unison; that is, either both switches are open, or they are both closed. FIGURE 7-4 illustrates how the CD4066 can be wired to simulate a DPDT switching function. Notice that all four of the bilateral switches contained in the device are used in this configuration.

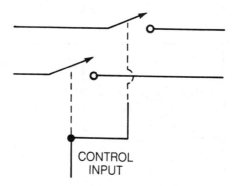

Fig. 7-3. The CD4066 wired to simulate a DPST switch.

The CD4066 features a very low amount of crosstalk between its on-chip switches. A typical isolation rating for this device is about 50 dB, which is almost as good as many mechanical switching circuits. The CD4066 is extremely fast. It can handle signals up to 30 MHz when it is powered off of a 12-volt source. The device's frequency response for switching on is 40 MHz. Noise immunity for this chip is typically 0.45 V_{DD}. When a switch is off, the leakage from input to output is typically 0.1 nA.

When a semiconductor switch is turned on, there is not a true short circuit between the input and the output, nor is there a true infinite resistance when a switch is off. Instead, turning one of the CD4066's switches on causes a drop from a very high to a low resistance. Although there is some variation from unit to unit, the CD4066 typically exhibits a resistance of about 80 ohms when on. This resistance holds fairly constant over the entire range of acceptable input signal voltages.

LIMITATIONS

A digital bilateral switch is a very powerful and versatile device, but it, like everything else, has some definite limitations and restrictions associated with its use. The voltage of the signal being switched must not be allowed to exceed the

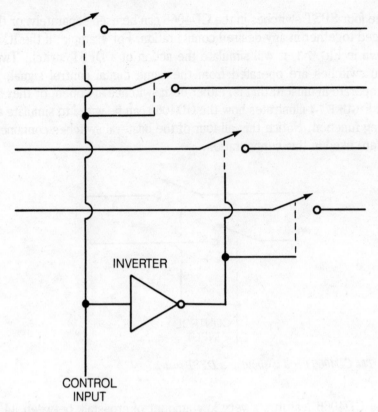

INVERTER

CONTROL
INPUT

Fig. 7-4. The CD4066 wired to simulate a DPDT switch.

chip's power supply voltage. Similarly, the voltage applied across the switch should not drop below ground potential, as seen by the bilateral switch IC.

Fortunately, CMOS ICs accept a fairly wide range of supply voltages. Anything up to 15 volts may be used. If you need to switch a signal that exceeds 15 volts, use an attenuator circuit (an amplifier with a gain of less than unity (one)) before the bilateral switch, and an amplifier at a later stage to bring the signal back up to the desired level.

If negative voltages must be switched by a bilateral switch, reference the signal to a floating, rather than true, ground. Another approach would be to use a bipolar voltage supply to power the chip. In this case, the IC would use a negative voltage as ground, while the input signal would be referenced to true ground (0 volts). If these voltage limitations are not taken into account in the circuit design, the bilateral switching unit is likely to be damaged or destroyed.

Similarly, the current of the signal to be switched also must be kept within specific limits to prevent damaging the bilateral switch IC. Different chips have different maximum current ratings. Check the manufacturer's specification sheet for the particular device you are planning to use. Typically, the current through a bilateral switch should be limited to a maximum of 15 mA (0.015 ampere) to 25 mA (0.025 ampere), depending on the specific unit being used.

CMOS ICs, including bilateral switches, are sensitive to static electricity. Newer devices have some on-chip protection, but it is best to take precautions to side-step possible problems.

Avoid touching the leads when the IC is out of the circuit. When storing a CMOS chip, it is advisable to short the leads together. There are various ways to do this. You could wrap the entire chip in aluminum foil. Alternatively, the ICs leads can be inserted into a special conductive foam, available from most electronics parts suppliers. Cases made of special anti-static plastic also may be used. Do not use ordinary plastic containers.

Never, under any circumstances, insert or remove a CMOS IC from a circuit that has power applied. This may be a temptation in some applications where IC sockets are used. Don't do it. The chip will almost certainly be ruined by such a procedure.

In designing a circuit using any CMOS devices, you must never leave any of the active inputs or outputs floating. That is, **all control terminals must be connected to something.** This is especially important for the digital control inputs of a bilateral switch. If a control input is left floating, the internal circuitry will be confused by the undefined digital state. At best, the circuit's operation may be unstable and erratic. More frequently, you will run into severe oscillation problems, which may go beyond annoying and actually do some damage to certain components in the circuit.

There are also practical limits to how fast a bilateral switch can be switched reliably between its on and off states. For most devices, the frequency response is somewhat dependent on the supply voltage used to power the IC.

THE CD4016 QUAD BILATERAL SWITCH IC

The CD4016 is another bilateral switch IC containing four independent digitally controlled SPST switches. The pin-out diagram for this chip is shown in FIG. 7-5. Compare this diagram with the pin-out diagram of the CD4066, shown back in FIG. 7-2. As you can see, these two devices are compatible pin for pin. For most applications, the CD4016 and the CD4066 are identical.

The CD4016 is an earlier version of the CD4066, and it offers no particular advantages of its own. Its specifications are slightly inferior to those of the CD4066. While the CD4066 is rated for an operating speed of up to 30 MHz at 12 volts, the CD4016's operating speed is limited to 25 MHz at 12 volts.

The biggest difference between these two chips lies in the switch resistance in the on state. The CD4066 has a typical on resistance of about 80 ohms. The CD4016's on resistance is typically about 300 ohms. For most noncritical appli-

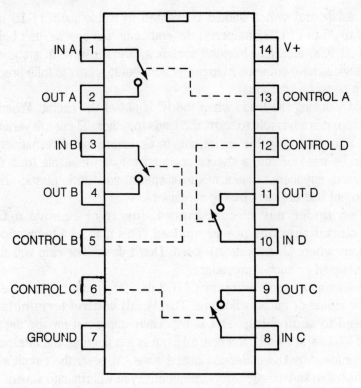

Fig. 7-5. The CD4016 quad bilateral switch IC is quite similar to the CD4066.

cations either the CD4016 or the CD4066 may be used without any changes in the circuitry. For more demanding applications, the CD4066 is definitely preferable.

The CD4016 is sliding into disuse, and you are likely to find it only on the surplus market these days. Still, you should be aware of this device, because sometimes it is available at very low prices.

APPLICATIONS FOR BILATERAL SWITCHES

Digitally controlled bilateral switches can be used in almost any standard automated switching application. They also can be employed in a number of less obvious applications. You will see a few unusual applications for bilateral switches in this section.

Programmable Resistor

A set of bilateral switches can be used as the heart of a digitally programmable resistor circuit, which can be used in place of any standard, two-terminal resistance element. A circuit of this type, built around a quad bilateral switch IC

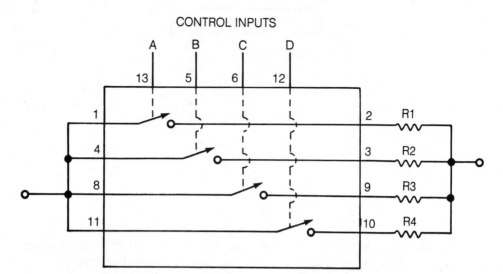

Fig. 7-6. The CD4066 used as a programmable resistor.

(like the CD4066) is illustrated in FIG. 7-6. A simple external digital control circuit can be used to select from among four fixed resistors (R1 through R4).

Two or more of the resistors may be selected simultaneously. If multiple resistors are switched into the circuit, the total effective value will be equal to the parallel combination of the selected resistances. You should recall the formulas for resistors in parallel from basic electronics:

$$1/R_T = 1/R_1 + 1/R_2 + 1/R_3 \dots + 1/R_N$$

Or, when there are just two resistors in parallel:

$$R_T = (R_1 \times R_2) / (R_1 + R_2)$$

For convenience in this discussion, assume that an open (off) switch has true infinite resistance. Also, ignore the bilateral switch's inherent on resistance. Since there are four binary inputs (A through D), there are 16 possible combinations that may be selected:

A	B	C	D	Resistors
0	0	0	0	none
0	0	0	1	R4
0	0	1	0	R3
0	0	1	1	R3, R4
0	1	0	0	R2

A	B	C	D	Resistors
0	1	0	1	R2, R4
0	1	1	0	R2, R3
0	1	1	1	R2, R3, R4
1	0	0	0	R1
1	0	0	1	R1, R4
1	0	1	0	R1, R3
1	0	1	1	R1, R3, R4
1	1	0	0	R1, R2
1	1	0	1	R1, R2, R4
1	1	1	0	R1, R2, R3
1	1	1	1	R1, R2, R3, R4

To give you a better feel for how this digitally controlled programmable resistor circuit functions here is a quick example. Assume that the following resistor values are being used in the circuit:

R1	10KΩ	(10,000 ohms)
R2	4.7KΩ	(4700 ohms)
R3	2.2KΩ	(2200 ohms)
R4	1KΩ	(1000 ohms)

Using these resistor values, the effective resistance of the circuit for each of the possible switching combinations are as follows:

A	B	C	D	Resistors	Effective Resistance
0	0	0	0	none	infinity
0	0	0	1	R4	1000 ohms
0	0	1	0	R3	2200 ohms
0	0	1	1	R3, R4	688 ohms
0	1	0	0	R2	4700 ohms
0	1	0	1	R2, R4	825 ohms
0	1	1	0	R2, R3	1499 ohms
0	1	1	1	R2, R3, R4	600 ohms
1	0	0	0	R1	10000 ohms
1	0	0	1	R1, R4	909 ohms
1	0	1	0	R1, R3	1803 ohms
1	0	1	1	R1, R3, R4	643 ohms
1	1	0	0	R1, R2	3198 ohms
1	1	0	1	R1, R2, R4	762 ohms
1	1	1	0	R1, R2, R3	1303 ohms
1	1	1	1	R1, R2, R3, R4	566 ohms

Notice that the resistance values do not change in a linear sequence with the digital control inputs. This may be awkward in some applications, but the system is functional, simple, and inexpensive.

Programmable Gain Amplifier

Another application for a set of bilateral switches is shown in FIG. 7-7. This is a digitally programmable gain amplifier circuit. The four bit (A through D) digital control signal determines the gain of the amplifier. Essentially, the Programmable Resistor circuit of FIG. 7-6 is used as the feedback resistor in a standard inverting op amp circuit.

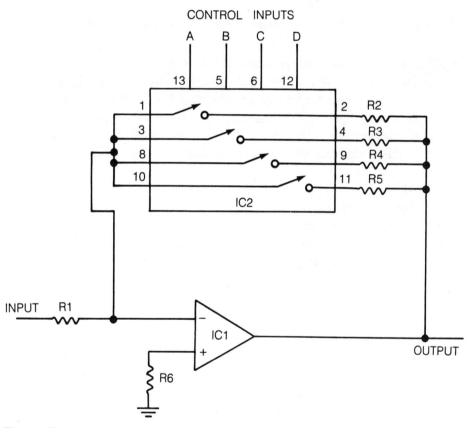

Fig. 7-7. This amplifier circuit features digitally controllable gain.

The basic inverting amplifier circuit is shown in FIG. 7-8 for comparison. Resistor Rf is the feedback resistance, and resistor R1 is the input resistance. The gain of the circuit is equal to:

$$A = -R_F/R_1$$

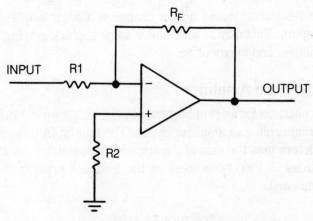

Fig. 7-8. The circuit of FIG. 7-7 is a variation on the basic inverting amplifier circuit.

The negative sign in this equation merely indicates polarity inversion. The output signal always will be 180 degrees out of phase with the input signal. If the input signal is positive, the output signal will be negative, and if the input signal is negative, then the output signal is positive.

By changing the value of the feedback resistance you can control the amplifier's gain. In this design example, assume the following resistor values:

R1	1K	(1000 ohms)
R2	10K	(10,000 ohms)
R3	4.7K	(4700 ohms)
R4	2.2K	(2200 ohms)
R5	1K	(1000 ohms)
R6	2.2K	(2200 ohms)

(The value of resistor R6 is not critical.)

Once again, since you have a four bit digital control input signal (A through D), there are sixteen possible switching combinations, which can be summarized as follows:

A	B	C	D	Effective rf Resistance	Gain
0	0	0	0	infinity	− infinity
0	0	0	1	1000 ohms	− 1
0	0	1	0	2200 ohms	− 2.2
0	0	1	1	688 ohms	− 0.69

A	B	C	D	Effective rf Resistance	Gain
0	1	0	0	4700 ohms	−4.7
0	1	0	1	825 ohms	−0.83
0	1	1	0	1499 ohms	−1.5
0	1	1	1	600 ohms	−0.6
1	0	0	0	10000 ohms	−10
1	0	0	1	909 ohms	−0.91
1	0	1	0	1803 ohms	−1.8
1	0	1	1	643 ohms	−0.64
1	1	0	0	3198 ohms	−3.2
1	1	0	1	762 ohms	−0.76
1	1	1	0	1303 ohms	−1.3
1	1	1	1	566 ohms	−0.57

Notice that the gain values do not change in a linear sequence with the digital control inputs. This may be awkward in some applications, but the system is functional, simple, and inexpensive.

Programmable Capacitor

The same technique used to create a digitally programmable resistor can also be used to set up a digitally programmable capacitor. The circuit is illustrated in FIG. 7-9. The four bilateral switches of a CD4066 IC are used to select

CONTROL INPUTS

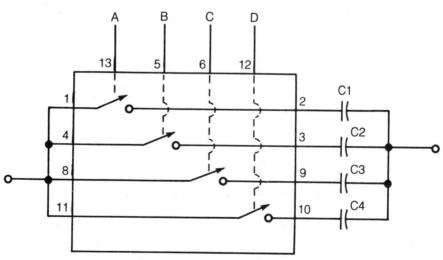

Fig. 7-9. The CD4066 can be used as a programmable capacitor.

various combinations of four capacitors. Since there are four digital input lines, there are 16 possible combinations.

When more than one capacitor is switched into the circuit at a time, the total effective capacitance will be equal to the parallel value of the selected capacitors. Capacitances in parallel simply are added together; that is:

$$C_T = C_1 + C_2 + C_3 \ldots + C_N$$

In a typical design of this digitally programmable capacitor circuit, the following capacitor values may be used:

C1   0.1 μF
C2   0.22 μF
C3   0.47 μF
C4   0.68 μF

If the circuit is made up of these capacitor values, the following combinations may be selected:

A	B	C	D	Capacitors	Effective Capacitance
0	0	0	0	none	---
0	0	0	1	C4	0.68 μF
0	0	1	0	C3	0.47 μF
0	0	1	1	C3, C4	1.15 μF
0	1	0	0	C2	0.22 μF
0	1	0	1	C2, C4	0.90 μF
0	1	1	0	C2, C3	0.69 μF
0	1	1	1	C2, C3, C4	1.37 μF
1	0	0	0	C1	0.10 μF
1	0	0	1	C1, C4	0.78 μF
1	0	1	0	C1, C3	0.57 μF
1	0	1	1	C1, C3, C4	1.25 μF
1	1	0	0	C1, C2	0.32 μF
1	1	0	1	C1, C2, C4	1.00 μF
1	1	1	0	C1, C2, C3	0.79 μF
1	1	1	1	C1, C2, C3, C4	1.47 μF

Notice that the total effective capacitance does not increase linearly with the digital control value. This system is slightly awkward in some applications, but it is still highly worthwhile for its low-cost and circuit simplicity.

Digitally Controlled Monostable Multivibrator

FIGURE 7-10 shows how the digitally programmable capacitor circuit of the preceding section can be used in a monostable multivibrator circuit. The delay time will be under digital control in this circuit. The multivibrator's timing capacitor is replaced with the programmable capacitor circuit of FIG. 7-9. The timing formula for this basic monostable multivibrator circuit, you should recall, is:

$$T = 1.1RC$$

In our design example for this circuit, assume the following component values:

C1	0.1 μF
C2	0.22 μF
C3	0.47 μF
C4	0.68 μF
R	220K (220,000 ohms)

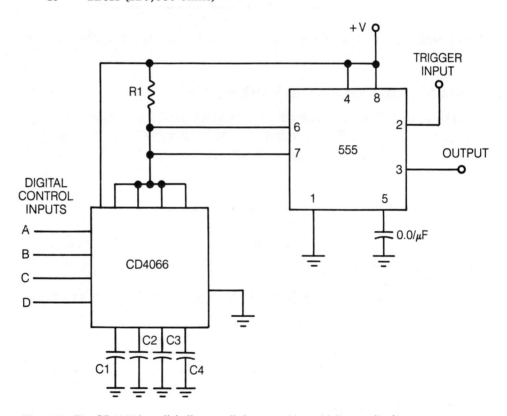

Fig. 7-10. The CD4066 in a digitally controlled monostable multivibrator circuit.

The sixteen possible combinations for this circuit are as follows:

A	B	C	D	Capacitance	Time
0	0	0	0		
0	0	0	1	0.68 μF	0.16 second
0	0	1	0	0.47 μF	0.11 second
0	0	1	1	1.15 μF	0.28 second
0	1	0	0	0.22 μF	0.06 second
0	1	0	1	0.90 μF	0.22 second
0	1	1	0	0.69 μF	0.17 second
0	1	1	1	1.37 μF	0.33 second
1	0	0	0	0.10 μF	0.02 second
1	0	0	1	0.78 μF	0.19 second
1	0	1	0	0.57 μF	0.14 second
1	0	1	1	1.25 μF	0.31 second
1	1	0	0	0.32 μF	0.08 second
1	1	0	1	1.00 μF	0.24 second
1	1	1	0	0.79 μF	0.19 second
1	1	1	1	1.47 μF	0.35 second

Notice that the timing period doesn't increase directly with the digital control values.

Digitally Controlled Astable Multivibrator

The same technique can be put to work in a digitally programmable astable multivibrator circuit, as illustrated in FIG. 7-11. For purposes of comparison, the basic astable multivibrator circuit is shown again in FIG. 7-12. The frequency equation for this circuit is:

$$F = 1.44/ ((R_A + 2R_B) C_T)$$

In our programmable astable multivibrator circuit, C_T is the total effective capacitance currently being selected via the programmable capacitor circuit (refer back to FIG. 7-9).

The component values in this design example of this circuit are as follows:

C1	0.1 μF
C2	0.22 μF
C3	0.47 μF
C4	0.68 μF
Ra	33K (33,000 ohms)
Rb	2.2K (2,200 ohms)

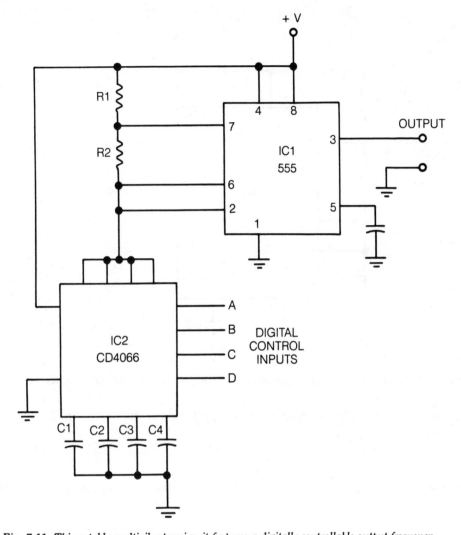

Fig. 7-11. This astable multivibrator circuit features a digitally controllable output frequency.

Since the resistor values (R_A and R_B) are constant in this circuit, you can partially solve the frequency equation before determining the various capacitance values:

$$F = 1.44/ (\,(R_A + 2R_B)C_T$$
$$= 1.44/ (\,(33000 + 2 \times 2200)\,)C_T$$
$$= 1.44/ (\,(33000 + 4400)C_T$$
$$= 1.44/ (35200 \times C_T)$$
$$= 0.000041/C_T$$

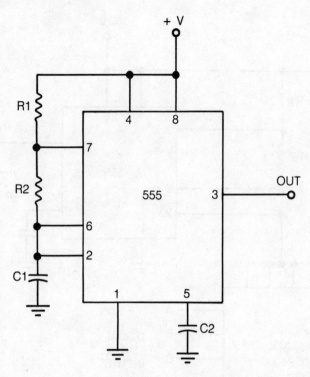

Fig. 7-12. The circuit of FIG 7-11 is a variation on this basic astable multivibrator circuit.

Or, if you convert the capacitance value from farads to microfards (μF):

$$F = 41/C_T$$

For each of the sixteen possible switching combinations, you get the following output frequencies:

A	B	C	D	Capacitance	Frequency
0	0	0	0		
0	0	0	1	0.68 μF	60.2 Hz
0	0	1	0	0.47 μF	87 Hz
0	0	1	1	1.15 μF	35.6 Hz
0	1	0	0	0.22 μF	185.9 Hz
0	1	0	1	0.90 μF	45.4 Hz
0	1	1	0	0.69 μF	59.3 Hz
0	1	1	1	1.37 μF	29.9 Hz
1	0	0	0	0.10 μF	409 Hz
1	0	0	1	0.78 μF	52.4 Hz
1	0	1	0	0.57 μF	71.2 Hz
1	0	1	1	1.25 μF	32.7 Hz

A	B	C	D	Capacitance	Frequency
1	1	0	0	0.32 μF	127.8 Hz
1	1	0	1	1.00 μF	40.9 Hz
1	1	1	0	0.79 μF	51.8 Hz
1	1	1	1	1.47 μF	27.8 Hz

As with the other circuits of this type presented in this section, the frequency does not increase directly with increases in the digital control values. Instead, you have a fairly complex and irregular pattern. This may complicate the control programming in some applications.

DIGITALLY CONTROLLED ROTARY SWITCHES

Closely related to the bilateral switch ICs we have been discussing so far in this chapter, are a series of CMOS chips which simulate the functioning of rotary switches. A fairly typical device of this type is the CD4051, which is illustrated in FIG. 7-13.

This IC is a one-of-eight switch, or a digital equivalent to a SP8T rotary switch. The control signals must be in proper CMOS digital form. (Three control bits are used.) The signal fed through the switch itself may be either digital or analog, providing that the voltage and current ratings for the IC are not exceeded.

Three control bits permit an unique control code for each of the switch's eight positions:

A	B	C	Position
0	0	0	0
0	0	1	1
0	1	0	2
0	1	1	3
1	0	0	4
1	0	1	5
1	1	0	6
1	1	1	7

Notice that the switch positions are counted from 0 to 7, instead of from 1 to 8. There are still eight steps because zero counts as an active position in this numbering scheme.

The common center pole of the switch is brought out to pin #3 of the IC. Although the switch contacts are not polarized, it is most convenient to consider

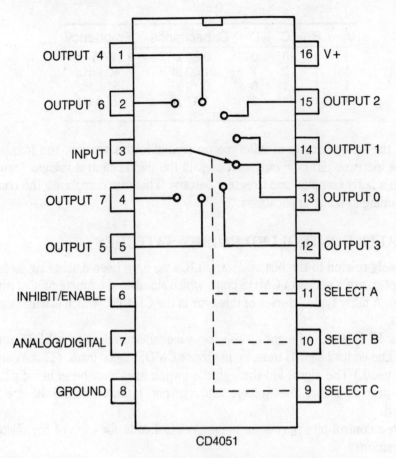

Fig. 7-13. The CD4051 is a CMOS which simulates a SP8T rotary switch.

this pin the input. The signal then can be fed to any one (and only one) of the eight output pins:

Output	Pin #
0	13
1	14
2	15
3	12
4	1
5	5
6	2
7	4

Three pins are used for the digital control signals input:

Control Input	Pin #
A	11
B	10
C	9

Pin #6 is an inhibit/enable terminal. If a high (logic 1) signal is applied to this pin, the switch input (pin #3) will be disconnected from all eight of the switch output pins. This function is included in the chip's circuitry, because in some applications it may be highly desirable to occasionally have no active channel connected in the circuit. Applying a low (logic 0) signal to pin #6 enables the CD4051. The input (pin #3) will be connected to the one output pin selected by the digital control inputs (A, B, and C).

Pin #7 permits the circuit designer to determine what types of signals will be switched by the chip. If this pin is grounded for a low (logic 0) signal, only digital signals may be passed through the switch. If you are going to be feeding analog signals through the CD4051's switch, then pin #7 should be connected to the most negative (lowest) voltage ever to be encountered in the system. In other words, this pin should be fed a constant voltage equal to (or possibly lower than) the most negative anticipated voltage that ever will be passed through the on-chip switch.

The analog voltages fed through the switch never should exceed the range defined by the chip's power supply. The positive voltage applied to pin #16 never should be exceeded, even momentarily. Similarly, the circuit designer should take precautions that the switch never will see a more negative voltage than the IC's ground potential connected to pin #8.

A typical value for the on resistance exhibited by the CD4051 is about 80 ohms, or about the same as the CD4066 quad bilateral switch IC.

FIGURE 7-14 shows a fairly typical application for the CD4051 one-of-eight switch. A three bit digital control value determines which of eight possible input signals will be fed through an output amplifier. Pin #6 can inhibit or enable the switch. In the inhibited mode, none of the input signals will reach the output amplifier.

Besides the obvious advantage of computerized automation, this circuit offers another important advantage over ordinary mechanical rotary switches, or similar devices. The various input signals do not have to be switched on and off in any particular order. You don't have to go through positions 2 and 3 to get from position 1 to position 4.

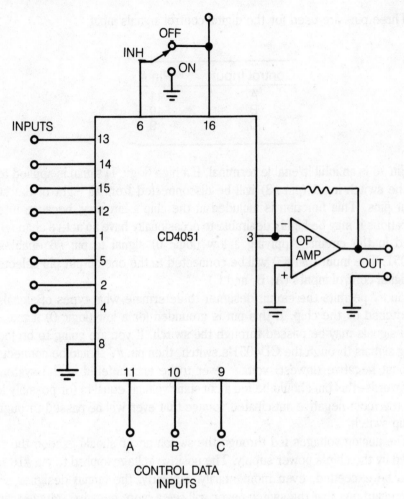

Fig. 7-14. The CD4051 can be used as an input selector for an amplifier.

There are a number of other digitally controlled multiple switch ICs available. All are basically similar to the CD4051. The CD4052 has 2 one-of-four switches, and the CD4053 features 3 one-of-two switches. One of the largest available digital switches is the CD4067, which simulates a SP16T, or one-of-sixteen switch. This chip's switching circuitry has a somewhat higher on resistance. A typical value is about 200 ohms. The CD4097 is similar to the CD4067, except this device features a pair of one-of-eight switches.

The CD4529 is a little unusual. Depending on how this chip is wired into the circuit, it can either simulate one SP8T switch, or a pair of SP4T switches. The on resistance for this device is fairly high. An on resistance of about 300 ohms or so is typical for the CD4529.

DC CONTROLLED AUDIO SWITCHES

A related type of IC intended for analog (audio) signal switching is the DC Controlled Audio Switch. Typical devices of this type are the LM1037 and the LM1038, both designed by National Semiconductor.

The LM1037, illustrated in FIG. 7-15, is used to select one of four stereo (two channel) input sources. The selected input is fed to a set of output lines. The output lines also can be muted; that is, all four of the available inputs may be disabled. The switching is controlled by applying a DC voltage level to the channel select control pins.

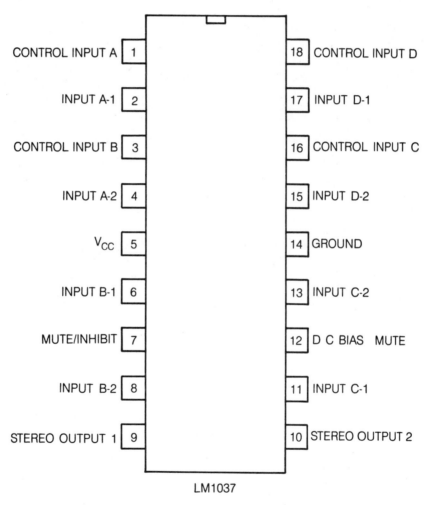

LM1037

Fig. 7-15. The LM1037 is used to select one of four stereo input sources.

The chip can handle audio signals up to 3.2 volts RMS. Crosstalk between the connected and unconnected channels is very low. A typical value for the crosstalk rating is −100 dB. The LM1038 is very similar to the LM1037, except this second chip's control inputs are designed to accept BCD (Binary Coded Decimal) signals.

8
Comparators

ANOTHER IMPORTANT TYPE OF SWITCHING CIRCUIT IS THE *COMPARATOR*. THIS is a circuit which switches its output in response to the relative values of two inputs. Comparators are fairly simple circuits, but they are quite powerful in terms of their potential applications.

In the simplest possible terms, a comparator is an electronic circuit which makes comparisons. The comparator is a close cousin to the popular op amp (operational amplifier). As a matter of fact, the same schematic symbol is used for both op amps and comparators, as shown in FIG. 8-1.

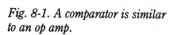 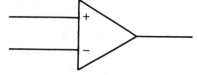

Fig. 8-1. A comparator is similar to an op amp.

Like the op amp, the comparator has two difference inputs—inverting and noninverting—and a single output. The signals at both inputs combine to control the signal at the output. An ordinary op amp can be easily wired to perform comparator functions. Dedicated comparators are specialized op amps.

The term *comparator* describes the function. A comparator is a circuit or device which compares two signals. Analog comparators work with signals in the form of voltages. Current comparators are not nearly as common as voltage comparators, but they are not unheard of.

A comparator looks at the two voltage signals at its inputs and determines which of the two signals is larger. This will be unambiguously indicated by the comparator's output signal. Most practical comparator circuits are quite sensitive. Even a difference of just a few millivolts between the difference inputs can trigger the output of the circuit. Generally speaking, for most applications, the more sensitive the comparator is, the better. The comparator function can be extremely handy in a great many applications, including A/D (analog-to-digital) converters, detecting high/low voltage limits, driving LED displays, metering, and even monostable and astable multivibrators.

SIMPLE COMPARATOR CIRCUITS

A simple comparator circuit is illustrated in FIG. 8-2. This circuit is as simple as it can get. There is nothing except for the comparator itself. No external components are required. Many practical comparator circuits will include some external components, such as resistors and capacitors, but we can reasonably ignore them for the time being.

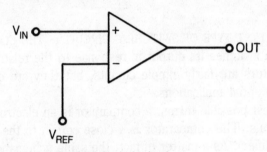

Fig. 8-2. The basic comparator circuit is extremely simple.

As in all comparator circuits there are two inputs: the inverting input and the noninverting input. The noninverting input is fed a known and fixed voltage signal called the *reference voltage*. The second unknown signal voltage is fed to the other (inverting) input and this is compared to the reference voltage. There are just three possibilities for the input signal. The unknown signal may be less than the reference voltage, equal to the reference voltage, or more than the reference voltage.

The operation of a comparator circuit can best be explained by looking at some specific examples. Let's assume the reference voltage in our comparator circuit has a value of two volts:

$$V_{REF} = 2 \text{ volts}$$

This is an arbitrary value selected only for demonstration purposes. Now consider each of the three possible comparator conditions. The unknown signal compared to the reference voltage is known as the input voltage, or V_{IN}.

While it is not quite accurate, for our purposes here we can assume that the comparator is essentially a difference amplifier with infinite gain. Since the non-inverting input voltage (V_{REF}) is held constant, the output voltage is controlled by the value of the inverting input voltage (V_{IN}) and its difference from the reference voltage.

The output voltage is equal to the difference of the two input voltages, multiplied by the gain:

$$V_O = (V_{REF} - V_{IN}) \times \text{Gain}$$

Remember that the comparator's gain, for this example is infinite, so the formula becomes:

$$V_O = (V_{REF} - V_{IN}) \times \infty$$

First, see what happens when the input voltage (V_{IN}) is less than the reference voltage:

$$V_{REF} = 2 \text{ volts}$$
$$V_{IN} = 1 \text{ volt}$$

$$
\begin{aligned}
V_O &= (2 - 1) \times \infty \\
&= 1 \times \infty \\
&= +\infty
\end{aligned}
$$

Any positive value multiplied by infinity will always be equal to positive infinity. No practical electronic device can put out an infinite voltage. The output will be clipped at the maximum positive voltage determined by the specific design of the circuit in question. In most cases, this maximum positive voltage will be just a little under the positive supply voltage value. As a result, when the input voltage (V_{IN}) is less than the reference voltage (V_{REF}), the output voltage will be essentially equal to $V+$.

Now, suppose the input voltage (V_{IN}) takes on a value higher than that of the reference voltage (V_{REF}):

$$V_{REF} = 2 \text{ volts}$$
$$V_{IN} = 3 \text{ volts}$$

$$V_0 = (2 - 3) \times \infty$$
$$= -1 \times \infty$$
$$= -\infty$$

Any negative value multiplied by infinity will be equal to negative infinity. Once again, no circuit can put out a truly infinite voltage. The output voltage again is limited to a specific maximum negative value, $-V$.

There is one other possible combination. What happens if the input voltage (V_{IN}) is exactly equal to the reference voltage (V_{REF})?:

$$V_{REF} = 2 \text{ volts}$$
$$V_{IN} = 2 \text{ volts}$$
$$V_0 = (2 - 2) \times \infty$$
$$= 0 \times \infty$$
$$= 0$$

Any value (including infinity) multiplied by zero equals zero.

All conceivable combinations of comparator inputs have been covered. The input voltage (V_{IN}) must be equal to, greater than, or less than the reference voltage (V_{REF}).

To summarize, the available combinations as follows:

Input voltage < reference voltage $V_0 = +V$
Input voltage = reference voltage $V_0 = 0$
Input voltage > reference voltage $V_0 = -V$

The response of a typical (ideal) comparator over its full operating range is illustrated in the graph of FIG. 8-3. Notice that the comparator circuit always gives a clear indication of whether the input voltage (V_{IN}) is greater than or less than (or equal to) the reference voltage (V_{REF}).

Certainly the capability of comparing voltages in this manner can be extremely useful in many practical applications. Some load device can be controlled or another circuit may take some specific action, depending on the comparative values of the input voltage (V_{IN}) and the reference voltage (V_{REF}).

LIMITATIONS OF PRACTICAL COMPARATORS

A practical circuit cannot have true infinite gain. There will be some inaccuracy if the input voltage (V_{IN}) is very close to, but not equal to, the reference voltage (V_{REF}). The comparator's output won't be able to switch states instanta-

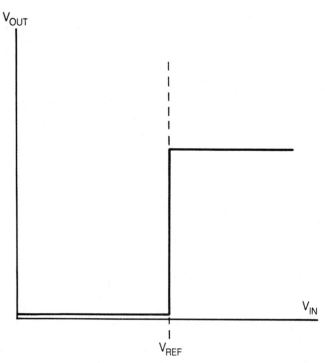

Fig. 8-3. The response graph for an ideal comparator.

neously, as shown in the graph of FIG. 8-3. Instead, there will be a narrow crossover region when there is only a small difference between the two input signals.

A typical operating graph for a practical comparator circuit is shown in FIG. 8-4. The crossover region is generally quite narrow. This usually will not cause any noticeable difference in the circuit's operation, except in some very high precision applications, or in applications involving very small input voltages.

When designing a comparator circuit, the reference voltage should be carefully selected. Any difference amplifier (op amp, or dedicated comparator) will have a maximum common-mode voltage rating. This is the maximum difference that is allowable between the two input signals. If this rating is exceeded, operation may become erratic, and the comparator IC could be damaged or destroyed.

As an example, say the comparator chip has a maximum common-mode voltage rating of up to 3.5 volts for its difference input voltages. If the reference voltage is 4 volts and the input voltage drops to zero, this limit will be exceeded and the device could be damaged or destroyed. Fortunately, this is a fairly extreme example. Most practical comparator devices have much larger maximum common-mode voltage ratings. In most applications, you will probably never have to worry about such problems. But you should be aware of the possibility for those rare instances when your application pushes the comparator to its extremes.

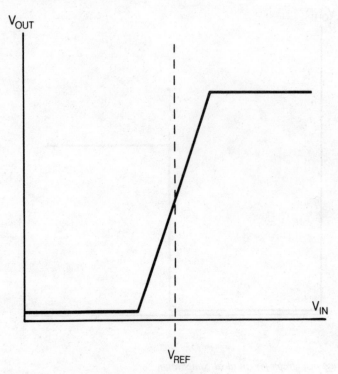

Fig. 8-4. The response graph for a practical comparator.

IMPROVED COMPARATOR CIRCUITS

While the basic comparator circuit discussed above will work, a practical comparator circuit normally uses a set of input resistors, as shown in FIG. 8-5. The function of these resistors is to limit the input current to a safe level. Aside from the current limiting of the inputs, this circuit functions in exactly the same way as the basic comparator circuit discussed above. The resistor values generally aren't critical, but for most practical applications, the two input resistors should have identical values.

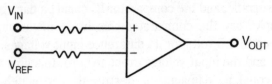

Fig. 8-5. Most practical comparator circuits include current limiting input resistors.

HYSTERESIS

A common problem with comparator circuits is output *chatter*. This problem can show up when the input voltage (V_{IN}) is very close in value to the reference voltage (V_{REF}). Under these conditions, the comparator's output may tend to oscillate between states. This oscillation is called *output chatter*. It is a result of the comparator's high gain. Unintended feedback through stray capacitances in the circuit can magnify chatter effects in some cases.

The problem can best be illustrated with an example. Imagine that the input signal to a comparator is a slowly rising voltage, with just a small amount of noise, as illustrated in FIG. 8-6. Through most of this signal's range, the comparator will work just fine. But when the input voltage is very close to the reference voltage, some problems may arise.

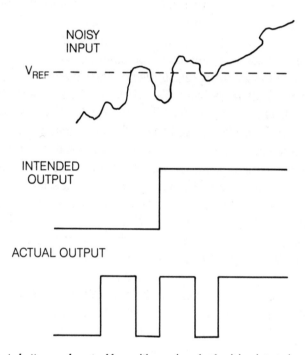

Fig. 8-6. Output chatter can be a problem with a noisy, slowly rising input signal.

As the input voltage slightly passes the reference voltage, the comparator's output goes high. If these two voltages are very close to one another, a little bit of signal noise or feedback, caused by a stray capacitance in the circuit, could cause the input voltage to momentarily drop below the reference voltage. This would force the comparator's output to switch back to its low state for a brief period, before returning to its true value.

In most comparator applications, chatter effects won't be frequent or severe, and won't be very important. In the majority of practical comparator applications, there are relatively large differences between the input voltage (V_{IN}) and the reference voltage (V_{REF}). A little chatter when V_{IN}'s value is close to that of

V_{REF} won't matter very much. Often, a little minor chattering can be reasonably ignored.

In applications where the input voltage (V_{IN}) spends a lot of time near the reference voltage (V_{REF}), output chatter could be a significant problem. Chatter will most likely be a problem in digital (or pseudo-digital) circuits, especially edge-triggered flip-flops, counters, and the like.

One good solution to output chatter problems is to add *hysteresis*, or an intentional lag effect, to the circuits. This usually is done with a feedback resistor. The effect of the feedback resistor is to make the turn-off level of the comparator somewhat lower than its turn-on level. This significantly reduces the tendency towards chatter. Another effect of the feedback resistor is to speed up the comparator's switching time. In many applications, the feedback resistor effectively can reduce noise components in the input signals.

The exact value of the feedback resistor is not particularly critical. Generally this resistor is given a fairly large value. Ten Megohm (10,000,000 ohms) resistors commonly are used for this purpose. The feedback resistor's value should be at least 1 Megohm (1,000,000 ohms) to obtain the benefits of hysteresis.

Just how does the feedback resistor create hysteresis in the circuit? A small percentage of the comparator's output voltage is fed back to its noninverting input through the resistor. When the output first switches into its high state ($V_{IN} > V_{REF}$), the feedback through the resistor will cause the voltage at the noninverting input to be slightly increased. The increase (feedback voltage) is greater than any anticipated noise signal. The feedback voltage cancels out the noise voltage and prevents the comparator's output from falsely switching back to low.

When hysteresis is used, the comparator's output can switch from low to high (or vice versa), with a nice, clean edge. The sharp transistion will not confuse any digital load circuitry. All but the most severe glitches are effectively filtered out of the comparator's output signal.

LOADING

To make any practical use of a comparator, or any other electronic circuit, it is necessary to feed the circuit's output signal into some load device, or circuit. This load could be almost anything, ranging from a simple LED indicator up to a full computer system. Any load will draw current from its source. If the load device, or circuit, attempts to draw more current than the source circuit (the comparator, in this case) can safely put out, either the source or the load (or often both) could be damaged. At the very best, the system will operate erratically.

In a poorly designed circuit, the load impedance could have an effect on the operation of the source circuitry. Output loading can be a problem in many comparator circuits, especially those built around general-purpose op amps. Most dedicated comparator ICs, however, feature built-in loading protection. This is true of the popular LM339 chip, which will be described shortly.

Similarly, the comparator itself, should not present too much of a load to it input sources. Fortunately, the majority of op amps and comparators draw very little input current, and represent negligible loads.

THE LM339 QUAD COMPARATOR

Many comparator circuits are built around standard, general-purpose op amp ICs. Such circuits are certainly functional, but they are, perhaps, not quite ideal because comparator circuits are just one of the many potential applications for op amps. To allow for all of the op amp's many applications, some compromises in design must be made.

The comparator function is so useful, that a number of dedicated comparator ICs have been developed. A dedicated device generally will offer better performance than a generic unit, because it has been designed specifically for comparator applications.

A number of dedicated comparator ICs are available today, but certainly the most popular of these is the LM339 quad comparator IC. This chip is so popular, it is being made by several different manufacturers. Some may use a different prefix in place of the LM but the type number still will be 339. The pin-out diagram for this popular 14-pin chip is illustrated in FIG. 8-7.

The LM339 contains four separate and electrically independent comparator stages in a single, compact package. The only pins shared by the individual comparator stages on this chip are the two power supply pins (pin #3 is V +, and pin #12 is ground, or V −). Notice that aside from the power supply pins, all of the pins on this device are either signal inputs or signal outputs. This suggests that the LM339 is very easy to work with, and it is.

One major advantage of the LM339 over the majority of standard op amp chips, is that a single-ended power supply may be used to drive the circuit. However, a dual-polarity power supply also may be used, if this suits the individual application. This power supply versatility is a great advantage over most op amp based comparators, which generally demand a dual-polarity power supply, regardless of the application.

There is nothing at all mysterious or exotic about the LM339 quad comparator. The comparator stages in this and similar comparator ICs are nothing more than slightly modified operational amplifier circuits, which have been customized for peak performance in comparator applications. Even though the LM339 is

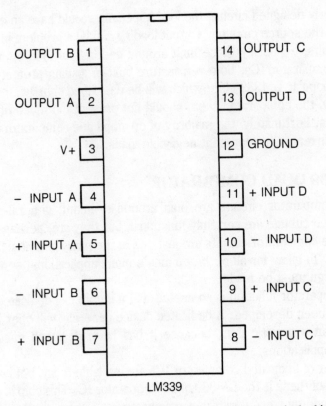

OUTPUT B	1		14	OUTPUT C
OUTPUT A	2		13	OUTPUT D
V+	3		12	GROUND
− INPUT A	4		11	+ INPUT D
+ INPUT A	5		10	− INPUT D
− INPUT B	6		9	+ INPUT C
+ INPUT B	7		8	− INPUT C

LM339

Fig. 8-7. The LM339 contains four independent comparator sections in a single chip.

designed for just one specific function (comparator), this chip can be used in countless different circuits. The basic comparator function is so useful, that the LM339 is a very versatile IC. The LM339 can also be used in a number of non-comparator applications.

Besides having so many potential applications, the LM339 also can accept many power supply voltages. This chip can be powered from a single-polarity power supply with a voltage of anywhere from +2 to +32 volts. If it is more appropriate to your specific application, the LM339 also can be operated from a dual-polarity power supply with voltages ranging from ±1 to ±18 volts.

The LM339 won't put much strain on even a light-duty power supply because the current drain is so minimal. Typically, this device draws only about 0.8 mA (0.0008 ampere), regardless of the supply voltage. The common-mode range of the LM339 comparator includes ground (0 volts), even if a single-polarity power supply is used for the circuit. This is not usually true for comparators built around op amps.

Now, take another look at the LM339's pin-out diagram (FIG. 8-7). It might occur to you that the arrangement of the pins on this device is a bit peculiar. The inputs and outputs don't appear to be very logically laid out at first glance. Actually, there is a very logical and important reason for the seeming oddity of the pin numbering system used on the LM339. If an output lead is positioned too near to the input leads, the comparator could break into uncontrollable oscillations. In some applications, it may be desirable to use a comparator as an oscillator, but that is certainly the exception, not the rule. In any case, anything uncontrollable is highly undesirable in any electronic circuit. To prevent such problems, the power supply pins are positioned to put a little distance between the outputs and the inputs.

The LM339 contains four comparator sections in a single package. This means that, in many applications, at least some of the comparator sections will be left unused. With some ICs, any unused sections can be simply left floating, with no connections to their input and output pins. But this is usually not advisable with the LM339. All unused pins of any comparator section on the chip should be hard-wired to ground. This is true for both inputs and outputs. If you're not using a pin, ground it.

One of the many convenient features of the LM339 is that the inputs can be driven from virtually any source impedance without loading effects. Each of the comparator outputs can sink up to 20 ma (0.02 ampere). The LM339's outputs are in the form of an open-collector NPN transistors for maximum circuit flexibility. An external pull-up resistor can be used with a different supply voltage than what is powering the rest of the chip. This makes the LM339 a good choice for interfacing analog circuits with any of the major logic (digital) families, including CMOS, DTL, ECL, MOS, and TTL.

The output voltage for the high state can be adjusted by the value of the pull-up resistor. The low state voltage is restricted to the comparator's negative supply voltage (ground or 0 volts if a single-polarity power supply is being used with the circuit).

9
Sample and Holds

IN THIS CHAPTER, WE WILL TAKE A LOOK AT A FAIRLY ESOTERIC TYPE OF SWITCH-ing circuit known as a *Sample and Hold*, or *S/H* circuit. The name sounds a little odd and mysterious, but it really spells out exactly what this type of circuit does. The input to a Sample and Hold circuit is a varying analog voltage. The circuit periodically takes a sample, or instantaneous measurement of the input voltage. This sampled value is held at the output until the next sample is taken.

SIGNAL SAMPLING

In most Sample and Hold applications, samples are taken many times per second. The greater the number of samples per cycle, the more the output signal will resemble the original input signal. FIGURE 9-1 shows the results of relatively infrequent samples. Compare this with the results of a higher sampling rate in FIG. 9-2.

If the sampling frequency is lower than the signal frequency, a phenomena known as *aliasing* will show up. The output will resemble a lower input frequency. It will be impossible to distinguish the sampled output of the too high input frequency from the sampled output of the lower (alias) frequency.

If the original signal frequency is to be recreated, the sampling rate must be higher than the highest anticipated input frequency. Theoretically, the sampling frequency must be at least twice the frequency of the highest possible input signal.

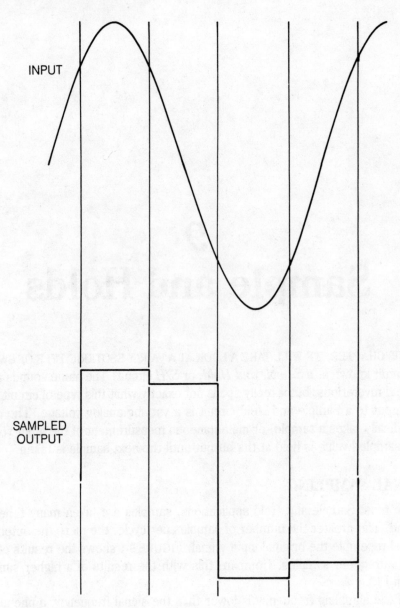

INPUT

SAMPLED
OUTPUT

Fig. 9-1. Infrequent samples result in an inaccurate reproduction of the original input signal.

APPLICATIONS

Sample and hold circuits have a number of practical applications. They are used most frequently in analog to digital (A/D) conversion systems. The stepped sample voltages can be converted more easily into a digital signal and recorded or transmitted than the original analog (varying) voltage signal. A digital to analog

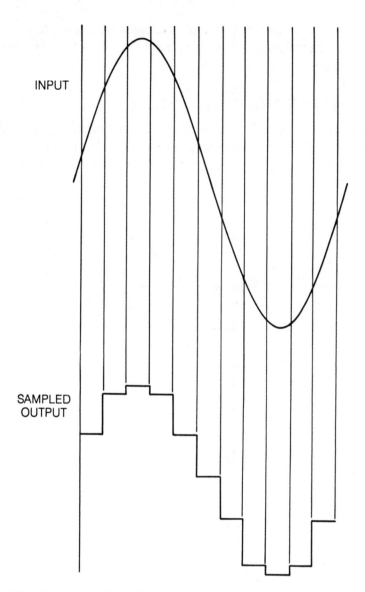

Fig. 9-2. More frequent samples result in a fairly accurate reproduction of the original input signal.

(D/A) converter is used at the system's output to recreate the original input signal. The more frequent the samples, the more accurate the reproduction will be.

The Sample and Hold circuit also finds applications in electronic music synthesizers, especially the older analog types. A Sample and Hold can be used to create complex waveforms, or, if a very low sampling frequency is used, a pattern of control voltages. If the sample frequency is related harmonically to the input frequency, the pattern will repeat noticeably. If the two frequencies are not related harmonically, the pattern of sampled voltages will appear to be random. Actually, the pattern will repeat eventually, but it will be so long, the repetition won't be recognized.

A SIMPLE SAMPLE AND HOLD CIRCUIT

FIGURE 9-3 shows a simple Sample and Hold demonstration circuit. In this demonstration circuit, the sampling function is controlled manually, via a manual switch. In most practical applications, an electronically controlled switching circuit will be used.

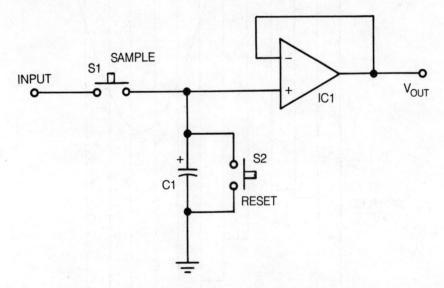

Fig. 9-3. This simple circuit demonstrates the functioning of a Sample and Hold.

Both control switches in this circuit (S1 and S2) are normally open (N.O.) SPST pushbutton switches. When switch S1 is momentarily closed, the circuit samples the instantaneous signal voltage at the input. This voltage is stored in capacitor C1. When the sampling switch is released, the stored voltage remains in the capacitor.

The op amp (IC1) is a buffer amplifier. I strongly recommend a high-grade type op amp for this application. Choose an op amp with a very high input impedance to minimize loading down the storage capacitor. Op amps with FET inputs usually do the best job in this type of application.

As long as power is applied to the circuit, the stored voltage on capacitor C1 theoretically will be held indefinitely. In practice, however, the capacitor's charge gradually will leak away. This drift effect can be minimized by using a high-grade, low-leakage capacitor, and an op amp with the highest possible input impedance.

The second switch in this circuit (S2) is used to reset the capacitor for the next sample. Briefly closing this switch shorts out the capacitor to ground, causing it to discharge very rapidly. The circuit then will be ready to take a new sample of the input signal. Be careful never to close simultaneously both the sample switch and the reset switch.

10
Switching Projects

IN THIS FINAL CHAPTER, YOU WILL LOOK AT A FEW SIMPLE, BUT INTRIGUING practical switching circuits for you to build and experiment with. Any of these projects can be constructed in an hour or two, at minimal cost. All of the required components should be fairly easy to find.

TOUCH SWITCH

With a touch switch, you can control almost anything with just the lightest touch. A touch switch may be activated with a fingertip, or, if your hands are full, an elbow, or a foot would work just as well. If you were inclined, you could even operate it with your nose. A touch switch circuit also can be used in many hidden sensor (alarm) applications. Aside from the convenience and practical applications, the light touch operation of a touch switch is fun and fascinating in its own right.

A simple, but practical touch switch circuit is shown in FIG. 10-1. The touch plate in this project is simply an exposed metallic (conductive) contact. A small piece of unetched copper-clad PC board will do a good job as the touch plate. The plate's dimensions will be influenced by the desired application. For most purposes, the plate probably should be about one to one and a half inches square.

It is extremely important to remember that *for safety, the touch plate must be 100 percent isolated from any ac power source*. The person operating this circuit comes into direct, uninsulated contact with a conductor (the touch plate) that is

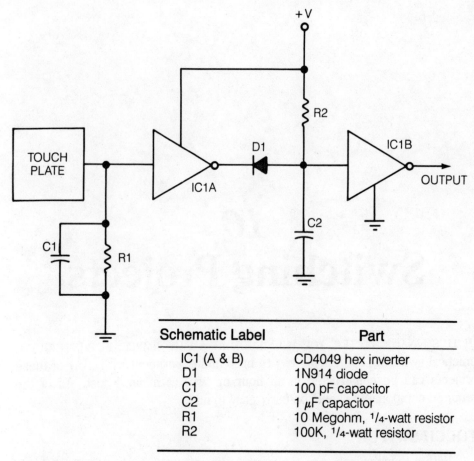

Schematic Label	Part
IC1 (A & B)	CD4049 hex inverter
D1	1N914 diode
C1	100 pF capacitor
C2	1 μF capacitor
R1	10 Megohm, 1/4-watt resistor
R2	100K, 1/4-watt resistor

Fig. 10-1. A touch switch is very easy to operate.

wired into the circuit. If any ac voltage manages to get through to the touch plate, the operator could suffer a painful, and quite possibly dangerous electrical shock. *Only low power dc voltages should ever be allowed to flow through any touch switch circuit.* Isolate the control circuit from any ac powered load through a relay, or an optoisolator. *The touch switch circuit itself should be operated off of battery power only.* An ac-to-dc converter type power supply conceivably could fail and feed 120 volts ac into the circuit and the touch plate. It is better to be safe than sorry. Don't take foolish and unnecessary chances.

I very strongly recommend that you breadboard this circuit before constructing a permanent version of the project. While the component values are not especially critical overall, sometimes this type of circuit may turn out to be a bit fussy. You might need to make some minor changes, or do a little fine tuning before the project operates reliably. In some cases, you might want to replace the two fixed resistors with trimpots to permit fine tuning of the circuit.

This circuit, and most other touch switch circuits, take advantage of the fact that 60 Hz power signals are almost always nearby. Low level 60 Hz signals from power lines are picked up by the operator's body, which acts as an antenna. The picked up signals are transmitted through a fingertip (or other body part) to the small touch plate contact. This input signal triggers the control circuit built around the two digital inverter stages, and the circuit's output goes high. When no contact is made with the touch plate, the circuit's output signal is in the low state.

TIMED TOUCH SWITCH

This project takes the idea of a touch switch circuit one step further, by combining it with a timer (or monostable multivibrator). The schematic diagram for this project appears in FIG. 10-2.

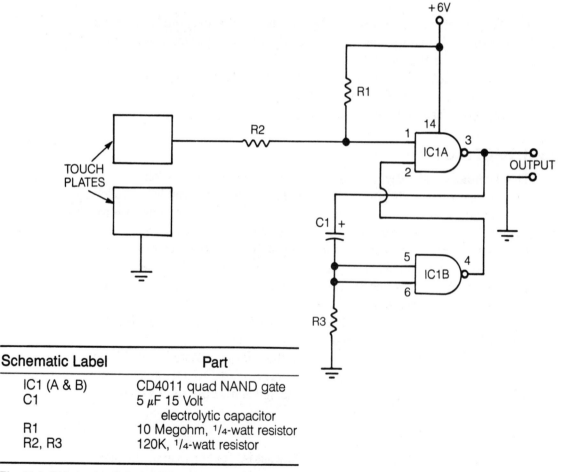

Schematic Label	Part
IC1 (A & B)	CD4011 quad NAND gate
C1	5 μF 15 Volt electrolytic capacitor
R1	10 Megohm, 1/4-watt resistor
R2, R3	120K, 1/4-watt resistor

Fig. 10-2. This touch switch circuit features a built-in time delay.

Normally, the output of this circuit is low. The circuit is activated by shorting a pair of touchplates with your finger, as illustrated in FIG. 10-3. When this is done the circuit's output will go high for a specific period of time (determined by certain component values), and then go low again, even if the touch plates are being shorted continuously.

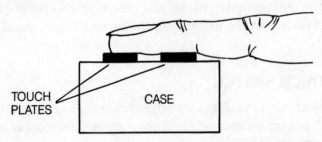

TOUCH
PLATES

CASE

Fig. 10-3. The circuit of FIG. 10-2 is activated by bridging the two touch plates with a fingertip.

For the component values given in the parts list of FIG. 10-2, the time period will be approximately one second. You are encouraged to experiment with other component values for this project.

As with any touch switch circuit, this project should be powered from batteries only. Refer back to the text for the preceeding project for more information on safety precautions for touch switch circuits.

LIGHT-ACTIVATED RELAY

This light-activated relay project permits fully automated (on/off) control over virtually any electrically powered device. This switching circuit is controlled by ambient light levels. The most obvious application for this circuit is as an automatic night light, which turns itself on at dusk, and off at dawn. The circuit, which is shown in FIG. 10-4, features a fully adjustable sensitivity control.

This project is not a terribly complex circuit. The op amp (IC1) is set up as a voltage comparator (see chapter 8). The voltage dropped across the sensor photoresistor (R1) is compared to the voltage dropped across the upper half of sensitivity potentiometer (R3). Adjusting the resistance of potentiometer R3 determines how much light is required to trigger the circuit.

When the comparator's output goes high, the relay is activated. Almost any external electrically powered device or circuit can be driven from the relay. Transistor Q1 is a small amplifier to boost the signal fed across the relay's coils.

SWITCH DEBOUNCER

No mechanical switch is perfect. After all, what is? Rather than neatly making or breaking contact (as shown in FIG. 10-5A) a mechanical switch tends to

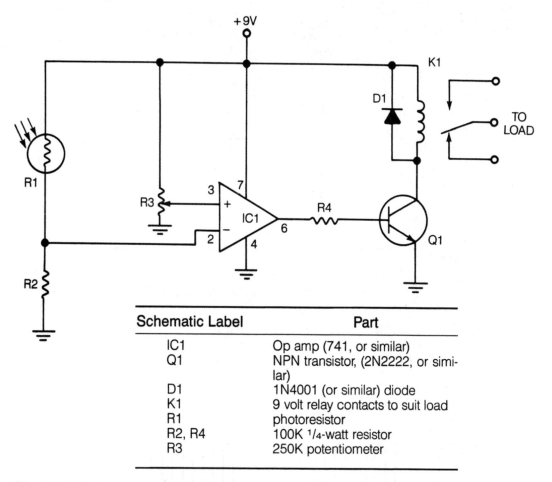

+9V

D1

K1

TO LOAD

R1

R3

3

7

+

IC1

R4

6

2

−

4

Q1

R2

Schematic Label	Part
IC1	Op amp (741, or similar)
Q1	NPN transistor, (2N2222, or similar)
D1	1N4001 (or similar) diode
K1	9 volt relay contacts to suit load
R1	photoresistor
R2, R4	100K ¼-watt resistor
R3	250K potentiometer

Fig. 10-4. This circuit turns itself on when the ambient lighting level drops below a specific, preset intensity.

bounce open and shut several times before settling into position (as shown in FIG. 10-5B). This bouncing takes only a tiny fraction of a second. For most analog applications, the switch bouncing rarely matters at all. It has no noticeable effect on the operation of the circuit.

Most digital circuits, however, are designed to recognize very brief pulses, so bouncing switch contacts could be a problem. For example, assume you want to increment a counter each time a pushbutton switch is depressed. When the switch bounces, it could cause the counter to be incremented several times, instead of just once per button push. Obviously, this could be a major problem, possibly even rendering some digital circuits completely useless.

What you need is some way to get the digital circuitry to ignore the unwanted bouncing contact pulses. A good way to do that is to have the first closure (first bounce) of the switch contacts trigger a monostable multivibrator with

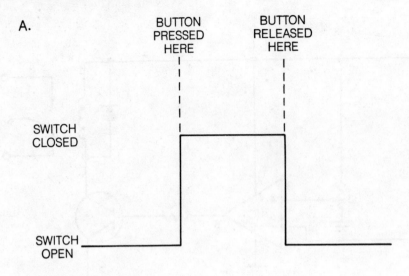

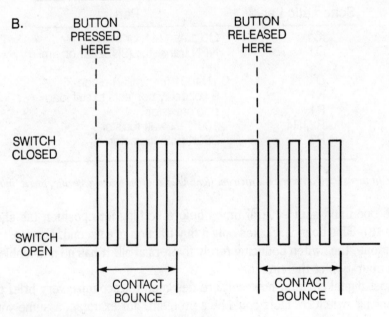

Fig. 10-5. Mechanical switches are prone to bouncing.

a time period that is slightly longer than the bouncing (settle in) time of the switch. The circuit for such a switch debouncer is shown in FIG. 10-6.

As soon as the switch contacts first make contact, the monostable multivibrator is triggered. Further switch openings and closing (bounces) will be ignored until the multivibrator completes its timing cycle. This timing period

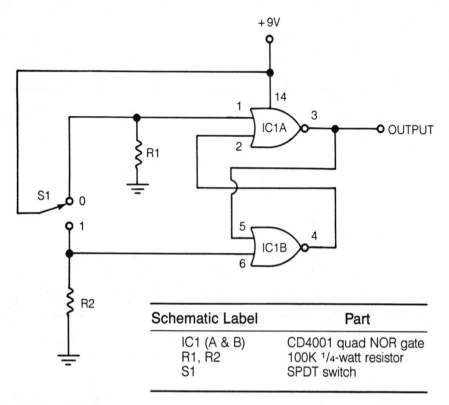

Fig. 10-6. A switch debouncer circuit.

does not have to be long by human standards. A fraction of a second is quite sufficient. The operation of this switch debouncer project is illustrated in FIG. 10-7.

Bounce-free switches will not be necessary for every manually operated switch in every digital circuit. When the need does arise, this simple project can make the difference between a functional high-technology piece of equipment and a useless piece of junk.

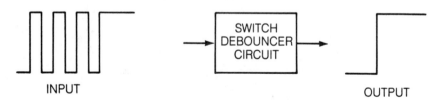

INPUT SWITCH DEBOUNCER CIRCUIT OUTPUT

Fig. 10-7. The circuit of FIG. 10-6 effectively blocks out any bouncing of the switch contacts.

FOUR-STEP SEQUENCER

The switching circuit shown in FIG. 10-8 is a four step sequencer. Four separate devices are switched on and off automatically, in sequence.

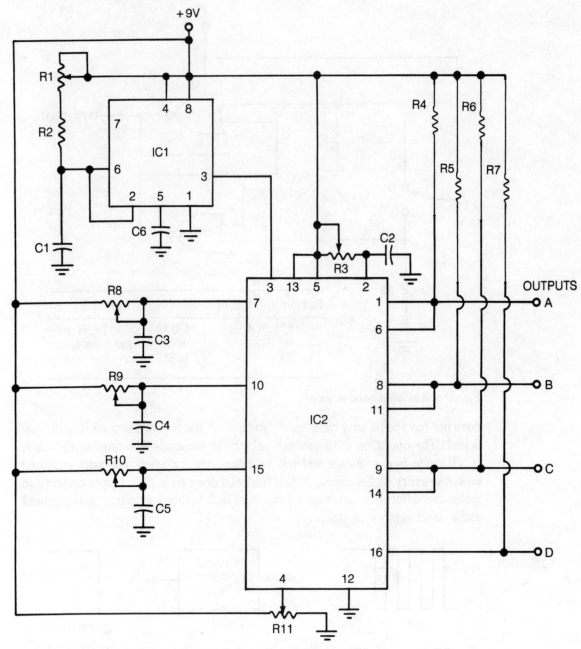

Fig. 10-8. This circuit sequentially switches between four separate output devices.

Component	Part
IC1	555 timer
IC2	558 quad timer
C1	0.0047-μF capacitor
C2-C5	0.1-μF capacitor
C6	0.01-μF capacitor
R1	500k potentiometer
R2	2.2k, $^1/_4$-watt resistor
R3, R8-R11	10k potentiometer (or trim-pot)
R4-R7	3.3k, $^1/_4$-watt resistor
R12	

Fig. 10-8. (Cont.)

The circuit uses five timers, in just two ICs. Four of the timers, set up as cascaded monostable multivibrators, are contained in a 558 quad timer IC. The fifth timer in this circuit is a separate 555 timer IC, in the astable mode.

In operation, one, and only one, of the four outputs is activated (high) at any time. First output A is turned on. When A is turned off, output B is turned on. Output C is next turned on, after B has timed out. When output C goes low, then output D is turned on. After output D times out, output A is activated again, and the entire cycle is repeated.

The on time for each output is individually adjustable via potentiometers R3, R8, R9, and R10. Potentiometer R1 and R11 interact to set the overall sequence rate.

Almost any circuit can be controlled with this project by using the outputs to drive suitable relays or SCR circuits. The potential applications of this project are limited only by your imagination.

DELAYED TRIGGER TIMER

Timer, or monostable multivibrator circuits are very useful in many control applications. The timer's output is normally low. When a trigger pulse is received, the output immediately goes high, and remains in that state for a fixed period of time, before reverting to the normal low condition. (In some circuits, the low and high states are reversed.)

In some specialized applications, you might not want the timing cycle to begin as soon as the trigger pulse is received. Instead, you might want the timing cycle to begin at some specific time after the trigger signal. In other words, the trigger effectively is delayed.

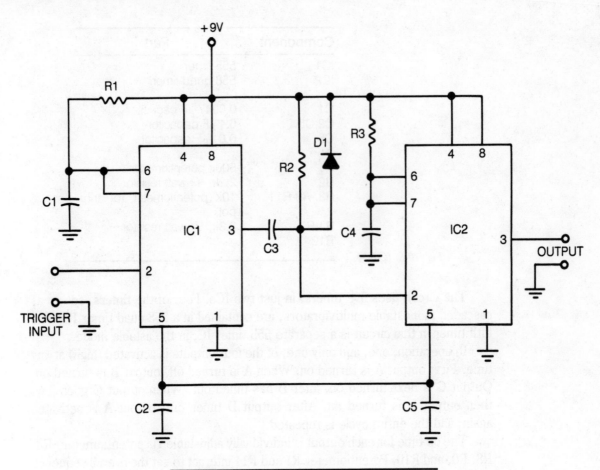

Component	Part
IC1, IC2	555 timer
D1	1N914 diode
C1	10-μF, 25-volt electrolytic capacitor*
C2, C5	0.01-μF capacitor
C3	0.001-μF capacitor
C4	0.5-μF capacitor*
R1	470k, $^1/_4$-watt resistor*
R2	10k, $^1/_4$-watt resistor
R3	820k, $^1/_4$-watt resistor*

*Timing component—experiment—see text

Fig. 10-9. This is a delayed trigger timer circuit.

The circuit for a delayed trigger timer project is shown in FIG. 10-9. Notice that this circuit is built around two 555 timers, each in the monostable mode. Of course, a single 556 dual timer IC could be substituted in place of the separate 555 ICs. Be sure to double-check the pin numbering if you make such a substitution.

In this circuit, IC1, and its associated components, control the delay time (T1) between the actual input pulse, and the beginning of the output pulse. IC2, and its associated components, control the length of the actual output pulse (T2). The operation of this circuit is illustrated in FIG. 10-10.

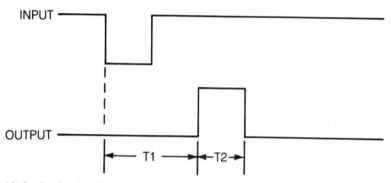

Fig. 10-10. In the circuit of FIG. 10-9, there is a delay between the trigger input pulse and the beginning of the output pulse.

The delay time (T1) for this circuit is set up by the values of resistor R1 and capacitor C1. The normal 555 monostable equation is used:

$$T_1 = 1.1 \times R_1 \times C_1$$

Similarly, the output pulse time (T2) is determined by the values of resistor R3 and capacitor C4:

$$T_2 = 1.1 \times R_3 \times C_4$$

Both of these times are fully independent of one another.

If you use the component values included in the parts list of Fig. 10-9, the delay period will be 5.17 seconds, and the output pulse time is about 0.45 second. In other words, the circuit's output goes high 5.17 seconds after the trigger pulse is received. At 5.62 seconds after the trigger pulse, the circuit's output goes low again.

You are encouraged to try other component values to create alternate timing periods. The timing components are indicated in the parts lists with asterisks (*).

Index

Other Bestsellers of Related Interest

TROUBLESHOOTING AND REPAIRING VCRs—
Gordon McComb

It's estimated that 50% of all American households today have at least one VCR. *Newsweek* magazine reports that most service operations charge a minimum of $40 just to look at a machine, and in some areas there's a minimum repair charge of $95 *plus the cost of any parts*. Now this time- and money-saving sourcebook gives you complete schematics and step-by-step details on general up-keep and repair of home VCRs—from the simple cleaning and lubricating of parts, to troubleshooting power and circuitry problems. 336 pages, 300 illustrations. Book No. 2960, $17.95 paperback, $27.95 hardcover

ALARMS: 55 Electronic Projects and Circuits—
Charles D. Rakes

Make your home or business a safer place to live and work—for a price you can afford. Almost anything can be monitored by an electronic alarm circuit—from detecting overheating equipment to low fluid levels, from smoke in a room to an intruder at the window. This book is designed to show you the great variety of alarms that are available. There are step-by-step instructions, work-in-progress diagrams, and troubleshooting tips and advice for building each project. 160 pages, 150 illustrations. Book No. 2996, $12.95 paperback, $19.95 hardcover

500 ELECTRONIC IC CIRCUITS WITH PRACTICAL APPLICATIONS—James A. Whitson

More than just an electronics book that provides circuit schematics or step-by-step projects, this complete sourcebook provides both practical electronics circuits AND the additional information you need about specific components. You will be able to use this guide to improve your IC circuit-building skills as well as become more familiar with some of the popular ICs. 336 pages, 600 illustrations. Book No. 2920, $19.95 paperback, $29.95 hardcover

EXPERIMENTS WITH EPROMS—Dave Prochnow

One of the greatest versatilities in advanced circuit design is EPROM (Erasable-Programmable Read-Only-Memory) programming. Now, Dave Prochnow takes an in-depth look at these special integrated circuits (ICs) that can be user-programmed to perform specific applications in a microcomputer. Fifteen fascinating experiments are a special feature of this book that presents not only the technology but also explains the use of EPROMs. 230 pages, 241 illustrations. Book No. 2962, $17.95 paperback, $24.95 hardcover

101 SOLDERLESS BREADBOARDING PROJECTS—Delton T. Horn

Would you like to build your own electronic circuits but can't find projects that allow for creative experimentation? Want to do more than just duplicate someone else's ideas? In anticipation of your needs, Delton T. Horn has put together the ideal project *ideas* book! It gives you the option of customizing each project. With over 100 circuits and circuit variables, you can design and build practical, useful devices form scratch! 220 pages, 273 illustrations. Book No. 2985, $15.95 paperback, $24.95 hardcover

GALLIUM ARSENIDE IC TECHNOLOGY: Principles and Practice—Neil Sciater

Now you can explore this decade's most exciting breakthrough in integrated circuit fabrication technology! With this book, Neil Sciater offers you the key to being on the cutting edge of the fastest growing field in semiconductor technology. Here is an excellent, non-mathematical overview of gallium arsenide (GaAs) ICs: how they are manufactured and packaged, what benefits they provide, and what the future holds for these innovative devices. 272 pages, 153 illustrations. Book No. 3089, $26.95 hardcover only

TROUBLESHOOTING AND REPAIRING THE NEW PERSONAL COMPUTERS—Art Margolis

A treasury of time- and money-saving tips and techniques that show personal computer owners and service technicians how to troubleshoot and repair today's new 8- and 16-bit computers (including IBM PC/XT/AT and compatibles, the Macintosh, the Amiga, the Commodores, and other popular brands). Margolis examines the symptoms, describes the problem, and indicates which chips or circuits are most likely to be the source of the trouble. 416 pages, 351 illustrations. Book No. 2809, $19.95 paperback, $29.95 hardcover

LASER EXPERIMENTER'S HANDBOOK— 2nd Edition—Delton T. Horn

This book provides necessary background on the theory and history of lasers while offering a number of practical, simple experiments that use relatively low power and present minimum risks. From continuous- and pulsed-beam lasers to a laser target and an actual laser transmitter and receiver, this handbook gives you the groundwork and the projects you need to get you started. 176 pages, 98 illustrations. Book No. 3115, $13.95 paperback, $21.95 hardcover

HOW TO TEST ALMOST EVERYTHING ELECTRONIC—2nd Edition—Jack Darr and Delton T. Horn

Completely current and authoritative, this single-volume reference provides up-to-date answers to present-day electronics problems. This revised edition offers practical, easy-to-follow instructions for using electronic test equipment. Described are electronic tests and measurements—how to make them with various types of electronic test equipment, and most importantly, how to *interpret* the results. 180 pages, 138 illustrations. Book No. 2925, $9.95 paperback, $16.95 hardcover

CUSTOMIZE YOUR PHONE: 15 Electronic Projects—Steve Sokolowski

Practical, fun, phone enhancement projects that anyone can build. A melody ringer, an automatic recorder, and a telephone lock—these are just a few of the improvements you can add to make your everyday telephone more interesting and more useful. Steve Sokolowski explains the vbasics of constructing an electronic projects as well as the fundamentals that make your telephone work. While the projects are all rather simple and inexpensive (built for about $10 to $30 each) they are also very useful. 176 pages, 125 illustrations. Book No. 3054, $12.95 paperback, $19.95 hardcover

THE LASER COOKBOOK: 88 Practical Projects— Gordon McComb

The laser is one of the most important inventions to come along this half of the 20th Century. This book provides 88 laser-based projects that are geared toward the garage-shop tinkerer on a limited budget. The projects vary form experimenting with laser optics and constructing a laser optical bench to using lasers for light shows, gunnery practice, even beginning and advanced holography. 400 pages, 356 illustrations. Book No. 3090, $18.95 paperback, $25.95 hardcover

BASIC ELECTRONICS THEORY—3rd Edition— Delton T. Horn

"All the information needed for a basic understanding of almost any electronic device or circuit . . ." was how *Radio-Electronics* magazine described the previous edition of this now-classic sourcebook. This completely updated and expanded 3rd edition provides a resource tool that belongs in a prominent place on every electronics bookshelf. Packed with illustrations, schematics, projects, and experiments, it's a book you won't want to miss! 544 pages, 650 illustrations. Book No. 3195, $21.95 paperback, $28.95 hardcover

101 OPTOELECTRONIC PROJECTS—Delton T. Horn

Discover the broad range of practical applications for optoelectronic devices! Here's a storehouse of practical optoelectronic projects including: power circuits, control circuits, sound circuits, flasher circuits, display circuits, game circuits, and many other fascinating projects! This book offers you an opportunity to make a hands-on investigation of the practical potential of optoelectronic devices. 240 pages, 273 illustrations. Book No. 3205, $16.95 paperback, $24.95 hardcover

44 POWER SUPPLIES FOR YOUR ELECTRONIC PROJECTS—Robert J. Traister and Jonathan L. Mayo

Here's a sourcebook that will make an invaluable addition to your electronics bookshelf whether you're a beginning hobbyist looking for a practical introduction to power supply technology, with specific applications . . . or a technician in need of a quick reference to power supply circuitry. You'll find guidance in building 44 supply circuits as well as how to use breadboards, boards, or even printed circuits of your own design. 220 pages, 208 illustrations. Book No. 2922, $15.95 paperback, $24.95 hardcover

MAINTAINING AND REPAIRING VCRs—
2nd Edition—Robert L. Goodman

". . . of immense use . . . all the necessary background for learning the art of troubleshooting popular brands" said *Electronics for You* about the first edition of this indispensable VCR handbook. Revised and enlarged, this illustrated guide provides complete, professional guidance on troubleshooting and repairing VCRs form all the major manufacturers, including VHS and Betamax systems and color video camcoders. Includes tips on use of test equipment and servicing techniques plus case history problems and solutions. 352 pages, 427 illustrations. Book No. 3103, $17.95 paperback, $27.95 hardcover

49 BATTERY-POWERED TWO-IC PROJECTS—
Delton T. Horn

Using detailed diagrams, parts lists, and step-by-step instructions, Horn shows you how to build useful two-IC devices including: delayed trigger timer, power amplifier, light range detector, digital filter and other inexpensive projects. These projects are perfect for developing the skills and the confidence you need to tackle more involved projects. Even if you're already building robots and computers, you can certainly enjoy the relaxation of whipping together a quick and simple device. 126 pages, 125 illustrations. Book No. 3165, $10.95 paperback, $17.95 hardcover

---